China Insights

This book series collects and presents cutting-edge studies on various issues that have emerged during the process of China's social and economic transformation, and promotes a comprehensive understanding of the economic, political, cultural and religious aspects of contemporary China. It brings together academic endeavors by contemporary Chinese researchers in various social science and related fields that record, interpret and analyze social phenomena that are unique to Chinese society, its reforms and rapid transition. This series offers a key English-language resource for researchers and students in China studies and related subjects, as well as for general interest readers looking to better grasp today's China. The book series is a cooperation project between Springer and China Social Sciences Press of China.

Kaizhong Yang
Editor

The Construction of Ecological Civilization in China

中国社会科学出版社
CHINA SOCIAL SCIENCES PRESS

Springer

Editor
Kaizhong Yang
Research Institute for Eco-civilization
Chinese Academy of Social Sciences
Beijing, China

Translated by
Wen Pang
School of Tourism and Human Geography
Xi'an International Studies University
Xi'an, China

ISSN 2363-7579 ISSN 2363-7587 (electronic)
China Insights
ISBN 978-981-99-6514-4 ISBN 978-981-99-6515-1 (eBook)
https://doi.org/10.1007/978-981-99-6515-1

Jointly published with China Social Sciences Press
The print edition is not for sale in China (Mainland). Customers from China (Mainland) please order the print book from: China Social Sciences Press.

This Springer imprint is published by the registered company Springer Nature Singapore Pte Ltd.
The registered company address is: 152 Beach Road, #21-01/04 Gateway East, Singapore 189721, Singapore

Series Foreword

Since the Opium War, modern China has come under attack and been bullied for its backwardness; this cultural circumstance has given many Chinese people a psychological inferiority complex, as China has lagged behind other countries technologically, institutionally and culturally. Efforts to change the situation in which Western countries were strong but China was weak and to revitalize China needed to start with cultural criticism and culture renovation. Therefore, the Chinese people turned their eyes to the outside world and learned from Japan, Europe, the USA, and even Soviet Russia. We have always been overwhelmed by stress and anxiety and have had a burning desire to reverse the state of being bullied as a result of underdevelopment, poverty and weakness and to catch up with and surpass the Western powers. In pursuing the more than 100-year-old dream of building a powerful country and reviving China, we have focused on understanding and learning from others, but seldom, if ever, have others learned from and understood us. This has not greatly changed in the course of modernization since China's reform and opening up in 1978. The translation and introduction of many Western works in the 1980s and 1990s is a very good example. This is the history of the Chinese people's understanding of the relationship between China and the rest of the world since the beginning of modern times.

At the same time, in pursuing the dream of turning China into a powerful country and rejuvenating it through material (technological) criticism, institutional criticism and cultural criticism, the Chinese people have struggled to find a path that would make the country prosperous and the people strong while preventing the country from being ruined and the race from being destroyed. This path first represents a thought, a banner and a soul. The key issue has been what kind of thought, banner and soul can save the country, making it prosperous and the people strong. For more than 100 years, the Chinese people have constantly carried out experiments and attempts amidst humiliation, failure and anxiety. They have experienced failure in adopting advanced Western technology and thought on the basis of safeguarding China's feudal system and practicing a constitutional monarchy after the collapse of the Western capitalist political path and a great setback in worldwide socialism in the early 1990s. The Chinese people ultimately embarked on a path towards a successful revolution with national independence and liberation; in particular, they have adopted a path leading to the socialist modernization of China—a road towards socialism with Chinese characteristics—by combining the theoretical logic of scientific socialism with the historical logic of China's social development. After more than 30 years of reform and opening up, China's socialist market economy has rapidly developed; tremendous achievements have been made in economic, political, cultural and social constructions; comprehensive national strength, cultural soft power and international influence have substantially improved; and a great success has been achieved in socialism with Chinese characteristics. Although the latter project has not yet become full-fledged, its systems and institutions have basically taken shape. After more than 100 years of pursuing dreams, China is rising among the nations of the world with a greater degree of confidence in the path it has chosen, the theory it has adopted and the institutions it has created.

Meanwhile, we should be aware that given the long-standing cognition and cultural psychology of learning from Western countries, we seldom take the initiative in showcasing ourselves—historical China and current China in reality—to the world, though China has emerged as a great world power. Due to a deeply rooted view that "Western countries are strong and China is weak", developed through Western-Chinese cultural exchanges, Western people and nations seldom have a sense of Chinese history or the current developments in China, let alone an understanding of China's developmental path and such in-depth issues as the scientificity and effectiveness of China's theory and institutions or their unique value for and contributions to human civilization. As self-recognition is not displayed, the "China Collapse Theory", "China Threat Theory", "China State Capitalism" and other so-called theories coined by certain people with ulterior motives and differing political views have been widely spread.

During our development, based on "crossing the river by feeling the stones", we have paid attention to learning from Western countries, understanding the world and learning to know ourselves through Western experience and discourse but have neglected self-recognition and efforts to let others know us. When we strive to become part of the world in a more tolerant and friendly way, we are not objectively, truly understood. Therefore, we should describe the path to the success of socialism with Chinese characteristics, tell Chinese stories, disseminate Chinese experiences, use international expressions to show a real China to the world, and help people around the world realize that the Western manner of modernization is not the endpoint of human historical evolution and that socialism with Chinese characteristics is also a valuable treasure of human thought. This is undoubtedly a very important task for an academic cultural researcher with a sense of justice and responsibility.

In this connection, the Chinese Academy of Social Sciences organized its top-notch experts and scholars and several external experts to write the *China Insights* series. This series not only provides an overview of China's path, theories and institutions but also objectively describes China's current development in the areas of political institutions, human rights, the rule of law, the economic system, finance, social governance, social security, population policies, values, religious faith, ethnic policies, rural issues, urbanization, industrialization, ecology, ancient civilization, literature, art, etc., thus depicting China in a way that helps readers visualize these topics.

We hope that this series will help domestic readers more correctly understand the course of the more than 100 years of China's modernization and more rationally look at current difficulties, enhance the urgency for and national confidence in comprehensively intensifying reform, build a consensus on reform and development and gather strength in this regard, as well as deepen foreign readers' understanding of China, thus fostering a better international environment for China's development.

Beijing, China
January 2014

Jianying Zhao

Contents

Contributors

Jingjing Shan Research Institute for Eco-civilization, Chinese Academy of Social Sciences (CASS), Beijing, China

Guangyao Sheng Research Institute for Eco-civilization, Chinese Academy of Social Sciences (CASS), Beijing, China

Wei Lou Research Institute for Eco-civilization, Chinese Academy of Social Sciences (CASS), Beijing, China

Meng Li Research Institute for Eco-civilization, Chinese Academy of Social Sciences (CASS), Beijing, China

Xiang Yu Research Institute for Eco-civilization, Chinese Academy of Social Sciences (CASS), Beijing, China

Shouxian Zhu Research Institute for Eco-civilization, Chinese Academy of Social Sciences (CASS), Beijing, China

Chengliang Huang Research Center on Xi Jinping Thought on Ecological Civilization of CASS, Beijing, China

Hongbo Chen School of Applied Economics, University of Chinese Academy of Social Sciences, Beijing, China

Rongxing Guo School of Applied Economics, University of Chinese Academy of Social Sciences, Beijing, China

Minjun Shi School of Public Affairs, Zhejiang University, Hangzhou, China

Guoqing Li School of Ethnology and Sociology, Minzu University of China, Beijing, China

Wenying Peng Capital University of Economics and Business, Beijing, China

Qiang Li University of Science and Technology Beijing, Beijing, China

Chapter 1
Ecological Civilization: A New Form of Human Civilization

Chengliang Huang

While summarizing Chinese civilization and national wisdom, China has deeply reflected on the environmental problems resulting from the development of industrial civilization in the last three centuries and therefore advocates the construction of ecological civilization at this crucial historical juncture in the transformation of human civilization. Based on the understanding of, respect for, and protection of nature, China carries out the construction of ecological civilization, demonstrating its explorations and practices in achieving sustainable development.

1 The Motives Behind the Rise of Ecological Civilization

The ecological civilization China advocates is a summary of the wisdom as manifested in the long and profound history and civilization of China, a reflection on the ecological problems brought about by industrial civilization, and a re-examination of the national condition in China after achieving high-speed economic growth with the traditional model of industrial civilization.

1.1 Ecological Problems of Industrial Civilization

While bringing material prosperity, industrial civilization has also led to tremendous pressure on resources and the ecosystem, as well as severe damage to the environment. As it pursues and honors the conquest of nature, industrial civilization aims at the maximum output of material wealth through the exploitation of various nonrenewable mineral resources and fossil energy. The development of industrial civilization has

C. Huang (✉)
Research Center on Xi Jinping Thought on Ecological Civilization of CASS, Beijing, China

K. Yang (ed.), *The Construction of Ecological Civilization in China*, China Insights,
https://doi.org/10.1007/978-981-99-6515-1_1

pushed the consumption of resources and energy to the limit that Earth can barely tolerate damaging regional and global ecological environments in a systematic and overall way. Early in the mid-20th century, there were eight serious environmental pollution incidents around the world, known as the eight public nuisance incidents, which were due to the emergence and development of modern industries such as chemistry, smelting, and automobiles. With continuously increasing emissions of three industrial wastes (industrial wastewater, waste gases and residues), incidents of environmental pollution and damage occurred more frequently.

Since the latter half of the 20th century, the resource and ecological environmental crises caused by industrial civilization have been seen globally, as these countries in the upstream of the industrial chain have been objectively alleviating the ecological and environmental pressure they face and improving the development quality by moving high-polluting and high-emission industries to other countries through industrial transfer. Meanwhile, countries receiving such industries, especially developing countries, must afford the price of environmental pollution and destruction to meet their own needs for development and the demand for capital. In addition, despite the historical advancement of modern science and technology, especially eco-environmental protection technology, two typical models of "cascading transmission" and "industrialization of environmental governance" are still in place for pollution control. The so-called "cascading transmission" refers to the shift of emissions and pollution through industrial transfer from industrial civilization to countries and regions with relatively poor resource endowment and lower labor costs during the process of globalization, which is promoted by capital. In general, traditional high-polluting industries are mainly moving from urban to rural, from suburban to distant suburban, from the West to the East, and from the North to the South. The so-called "industrialization of environmental governance" means the overreliance on technical and economic power and the blind construction of vast environmental protection industries in industrialized countries and the control of pollution from one type of equipment with another, thus resulting in greater consumption of resources and more environmental pollution. Therefore, it is unable to address both the symptoms and the root causes of eco-environmental problems but rather causes sharper conflicts between humans and nature.

1.2 Rethinking of China on Traditional Industrial Civilization

Chinese people have always revered and loved nature, and China's 5000-year-long civilization embodies a rich ecological culture and a governance philosophy. The Communist Party of China (CPC) and the Chinese government have always attached great importance to ecological environmental protection and ecological civilization construction. In the early days after the founding of the People's Republic of China (PRC) in 1949, the first-generation leaders of PRC, represented by Mao Zedong, started the systematic water conservancy projects of the Huaihe River, Yangtze River, Yellow River and Haihe River, which made a mark in history. Chairman Mao Zedong also called for "greening the motherland" to make "everywhere beautiful in China". In June 1972, the Chinese government sent a delegation to participate in the United Nations Conference on the Human Environment, which was a landmark in the history of human development. In 1973, at the first National Environmental Protection Conference, the policy of "overall planning and reasonable layout, comprehensive utilization and turning wastes into resources, relying on the masses and encouraging social participation, protecting the environment for the benefit of the people" was proposed. At the 16th CPC National Congress, China set the goal of building a resource-saving and environmentally friendly society under the guidance of the scientific outlook on development, and the 18th CPC National Congress incorporated the construction of ecological civilization into the overall plan for building socialism with Chinese characteristics.

At the beginning, newly founded China had an extremely weak economic foundation, with an annual per capita income of less than USD 13.84. However, in a few decades, China has become the second largest economy in the world and completed the journey of industrialization that took developed countries for centuries. China truly created a miracle of rapid economic and social development but accumulated some ecological and environmental problems, mainly because the rapidly progressing industrialization in China brought up an industrial system with a relatively extensive production mode that was composed of thermal power, iron and steel, petrochemicals, cement, coking and other industries, resulting in high energy consumption, emissions and pollution. Besides, China was then the "factory of the world", producing and exporting labor-intensive and low-end products of the industrial chain. Besides, many high-polluting industries were moved to China, making the "polluter's paradise" effect obvious here. Clearly, it is up to the basic national conditions that China could not follow the traditional extensive development path, as a huge population, scarce cultivated land, limited resources, and insufficient capacity to tolerate environmental pollution are the basic national conditions China has to face in the long run.

Since the 18th CPC National Congress, the CPC and the Chinese government have been reflecting more deeply and systematically on the shortcomings of the traditional development model. In particular, in the course of formulating, enriching and developing the overall plan for building socialism with Chinese characteristics, China has incorporated the construction of ecological civilization into the "five-sphere integrated plan", that is, to promote coordinated progress in the economic, political, cultural, social and eco-environmental fields; the overall plan for building socialism with Chinese characteristics, and proposed to construct a modernization in which humans and nature coexist in harmony, making it a distinctive feature of China's road to Modernization.

2 Ecological Civilization System

The ecological civilization China advocates is based on the premise of grasping the laws of nature, respecting and preserving nature, and pursuing harmony between humans and nature, between individuals and between individuals and society. Ecological civilization construction in China is a comprehensive and systemic strategic project since building an ecological civilization system essentially serves the construction of a society that is characterized by the harmonious coexistence of humans and nature and maps out the path to an all-round green transformation in China based on economic, cultural, political and social development.

2.1 Ecological Cultural System Based on Ecological Values

The construction of ecological civilization focuses on equal treatment and humanistic care for nature in terms of values. Humans and nature coexist on Earth, and they are both independent and interdependent. Humans can only enjoy a sustainable supply of natural resources for a long time by exploring, protecting and developing nature on the condition of respecting the laws of nature and giving nature humanistic care. The construction of ecological civilization will educate society with ecological culture and consciousness, turning environmental ideas into popular cultural awareness and ecological ethics into social morals. The values of ecological civilization would accelerate the transformation of production and living modes from the traditional "conquest of nature" to the "harmonious coexistence of humans and nature" and from the pursuit of profit maximization to ecological welfare maximization. In short, the construction of an ecological civilization system proposes to cast away the nature governance idea of humans ruling over nature and anthropocentrism but to set up values of "respecting and protecting nature and obeying nature's laws" to realize the simultaneous ecological development of science, philosophy, morality, art, and religion, establish the values of harmonious development of humanity and nature and realize the common prosperity for humanity and nature.

2.2 An Economic System Aimed at Integrating Industries and Ecosystems

The practices of ecological civilization construction are manifested in the deep and wide integration between traditional industries and ecosystems and the normalized development of ecological industries. Ecological civilization is to abandon the production and living modes of plundering and conquering nature, explore new technologies and new energy sources, learn from the wisdom of nature, adopt ecological technologies and techniques, make comprehensive and rational use of natural resources, protect natural values and achieve green development, and build a modernization featuring harmonious coexistence of humans and nature, ensuring a "win–win" situation for both humans and nature. The construction of ecological civilization is reflected in self-conscious and self-disciplined production and living modes, the pursuit of a benign interaction between economy and ecology, ecological economic operation, changes in the high-resource input and high-pollution production methods, the virtuous cycle of social material production system with ecological technology, and a green industrial structure with environment- friendly industries becoming dominant and an important momentum to economic growth.

2.3 An Institutional Framework for Promoting Ecological Progress Guaranteed by a Modernized Governance System and Capacity

The institutional construction of ecological civilization targets establishing a systematic and complete institutional mechanism that is conducive to promoting and realizing the harmonious coexistence of humans and nature. Since the 18th CPC National Congress, China has made significant breakthroughs in comprehensively deepening reforms in ecological civilization construction and accelerating the establishment of a systematic and complete institutional framework. The CPC has strengthened its overall leadership on ecological civilization construction, made comprehensive efforts in fields including ideology, law, systems, organization and style, better perfected top-level design, and institutionally improved the leading role of the CPC in ecological civilization construction. In particular, China carries out the inspection system of the central government for ecological environmental protection to ensure that both CPC committees and governments are held responsible for protecting the ecological environment and that officials take responsibility for ecological environmental protection in performing their duties and provide guarantees with the ecological conservation performance assessment and accountability system for environmental protection and improvements in living. Centering on the construction of the ecological conservation performance assessment and accountability system, Hebei provincial government has made active explorations in "what to be assessed, and how to assess and define the responsibility", and integrated the ecological civilization construction with cadre performance assessment system by optimizing the subjects, indicators, the proportions of economic achievements and so on. The ecological conservation performance assessment has helped alter the extensive development model in some regions and departments that focused on GDP only, truly reflects the environmental pollution and ecological costs behind economic growth, exposes the problems caused by "focusing on production while ignoring ecology", and more comprehensively evaluates and measures the "quality" and "quantity" of development, while forming a scientific guiding and binding mechanism and boosting governmental officials at various levels to have a correct understanding of performance and outlook on development, ensuring that the economy and society develop along the right path. In 2021, the air quality in Hebei Province improved significantly, and the number of days with good air quality reached 269, meaning that the proportion of days with good air quality reached 70% for the first time on record.[1]

2.4 An Objective Appraisal System Centering on Improvements in Ecological Environmental Quality

To improve environmental quality, the construction of ecological civilization emphasizes fundamentally solving ecological and environmental problems by reducing the pressure from economic activities and human behavior on natural resources and the ecological environment to an extent tolerable and giving time and space for natural ecology to recuperate. The Chinese government has defined and enforced strict measures to ensure eco-environmental quality, including red lines for environmental protection, minimum standards for environmental quality, and ceilings on resource utilization. In terms of red lines for environmental protection China established a strict system of controls that covers important ecological spaces so that these spaces do not change in character, decrease or lose ecological functions. In terms of minimum standards for environmental quality, China makes it the basic requirement that environmental quality must not worsen and may only get better, striving for constant improvements on this basis, and strictly enforces accountability in areas with serious ecological degradation or where environmental quality has worsened. In terms of ceilings on resource utilization, China will consider not only the needs of people and the world today but also those of the environment and future generations by keeping a firm handle on the intensity of natural resource exploitation and utilization to ensure that the carrying capacity of these resources is not exceeded.

[1] "Significant Improvements Made in Environmental and Air Quality, Proportion of Days with Good Air Quality Topped 70% in 2021 for the First Time", *Hebei Daily*, http://hbepb.hebei.gov.cn/hbhjt/xwzx/meitibobao/101653476179686/html.

These strategic initiatives are mainly to improve the quality of the ecological environment and secure a solid foundation for ecological civilization construction. Across China, the governments at all levels have made efforts to resolutely control pollution and protect the environment. For example, China has implemented three major action plans for preventing and controlling air, water, and soil pollution and announced a battle against pollution to improve air, water and soil quality to concentrate on overcoming prominent environmental problems that impact the people directly, improving China's environmental quality significantly.

2.5 *Ecological Security System Focusing on Virtuous Cycle of Ecosystems and Effective Environmental Risk Prevention*

In terms of the national security system, the construction of ecological civilization puts ecological security on par with political security, military security and economic security as one of China's national security strategies. In 2004, in the *Law of the People's Republic of China on the Prevention and Control of Environmental Pollution by Solid Waste,* the Standing Committee of China's National People's Congress clarified the legislative purpose that the law is to prevent and control environmental pollution by solid waste, protect human health, maintain ecological security and promote sustainable economic and social development.

The CPC and the Chinese government have realized that it is necessary to build a national security system that integrates political security, homeland security, military security, economic security, cultural security, social security, scientific and technological security, information security, ecological security, resource security, and nuclear security to implement the holistic concept of national security. The *Outline of the Yellow River Basin's Ecological Protection and High-quality Development Plan,* which was issued in 2021, proposed to accelerate the construction of a solid and sound, well-supported national ecological security barrier, setting up a model for less developed and ecologically fragile areas for the construction of ecological civilization. The Yellow River Basin is an important ecological barrier in China, an ecological corridor connecting the Qinghai-Xizang Plateau, Chinese Loess Plateau and North China Plain, and is composed of a number of national parks and national key ecological function areas such as Sanjiangyuan (the source of the Yangtze, Yellow and Lancang Rivers) National Park and Qilian Mountain National Park. Aimed at both preventing major floods and disasters and coordinating development and security, China successively planned ecological protection and high-quality development strategies for the Yellow River Basin to strengthen risk prevention and response capabilities and accelerate the construction of the Yellow River Basin into a defense line against natural disasters to make the Yellow River strategy gain ground among people and ensure that the Yellow River is calm year after year. The construction of sound and well-supported ecological security in large river basins is an important element to significantly improve national ecological security.

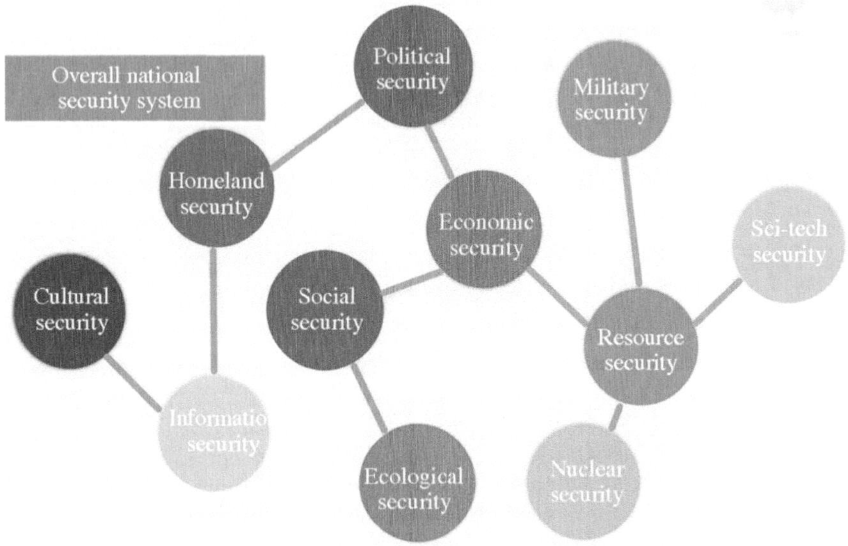

China's overall national security system

3 People-Centered Ecological Civilization

Looking back at history, the excessive pursuit of material wealth has caused great tension between humans and nature and a serious crisis of human survival. Meanwhile, extreme environmentalists emphasize the need for human society to completely stop activities to transform nature. However, the practical results show that neither has correctly understood or scientifically dealt with the relationship between humans and nature nor adapted to the development needs of human society. The construction of ecological civilization China advocates is essentially a matter of correctly handling the relationship between humans and nature, and the core idea is to form a value system that ensures equity between humans and nature, between individuals of the present generation, and between individuals of present and future generations. Ecological civilization upholds the people-centered development philosophy, placing people first, while opposing extreme anthropocentrism and extreme ecocentrism, stressing that people are the center of values but not the master of nature, that nature is the basis of values and the partner of people, and that the comprehensive development of people has to promote the harmonious coexistence between humans and nature.

3.1 The Core or Fundamental Value Standpoint of China's Ecological Civilization Construction is the People-Centered Development Philosophy

Industrial civilization advocates anthropocentrism and over-emphasizes the conquest of nature by humans, which has caused damage to the natural ecosystem and ecological environment, leading to frequent environmental problems and climate extremes and a huge challenge for the sustainable development of human society. Since the 1960s and 1970s, a number of new ecological schools of thought have emerged in Western societies, and anthropocentrism has become "weak", with the rise of biocentrism and ecocentrism advocated by nonanthropologists and panecologists. Some radical Western ecologists and social activists of the green movement even oppose human consumption of animal food to meet the basic needs for survival of human society and even believe that plants are also spirits that can talk and think, thus ignoring the value of nature as a whole for each other's existence and denying the fundamental value of human beings for nature and for social existence. Since ancient times, in China, there has been an ecological culture believing that humans and nature are equal and emphasizing both respecting, accommodating and protecting nature and adhering to the people-centered development philosophy.

China's ecological civilization is based on the people-centered development philosophy, requiring a re-examination of the values of industrial civilization that "man can conquer nature". The belief that only humans have value has been popular for centuries and has been deeply rooted in industrial civilization. For example, Kant proposed that "man is the end" and "man is the ultimate legislator of nature", which developed into science and philosophy that deny the value of nature. Industrial civilization played a positive role in promoting social and economic development at a specific period of history, but it is no longer able to solve current global challenges such as environmental problems, climate extremes and resource crises and cannot meet the demand for future development of human society toward a more advanced civilization.

China regards ecological civilization as the greatest well-being of the people and always takes the fundamental interests of the people as the highest criterion and ultimate measure of ecological civilization construction. It has been clarified that the construction of ecological Civilization will keep economic construction as the central task and create more material wealth to meet people's ever-growing needs for a better life in the era of high-quality development, meaning that people's needs are shifting from "survival" to "ecology" and from "subsistence" to "environmental protection" over time. The development of civilization is to provide more high-quality ecological products to continuously meet the growing needs of the people for a beautiful ecological environment.

3.2 *Insisting on People-Centered Ecological Civilization Construction with Rich Connotations*

Since the entry of a new era of social development, Chinese people have a growing need for a better living environment, the maintenance of environmental rights, and public ecological products, which is challenged by the limited carrying capacity of ecological resources and the environment, insufficient ecological public goods, and the increasingly serious ecological and environmental situation, so new understanding and new practices are needed to solve this contradiction. Objective facts have proven that a good ecological environment is the most equitable public good and the most universally beneficial welfare.

The people-centered philosophy deepens and expands the concept of people's livelihood. The ecological environment provides elementary living conditions for the people and basic elements for social production, and it concerns the fundamental interests of the greatest majority of people. All citizens will benefit from the well-protected ecological environment, and the development of the whole society will stagnate if the ecological environment is damaged. The ecological environment is obviously universal and equitable and has typical properties of public goods. As the eco-environmental problems become increasingly serious and the impact on social life grows rapidly, the public goods properties of the ecological environment become increasingly obvious, and as special public goods, the ecological environment is the most important of all.

Qinghai Province is steadily advancing the construction of the eco-environment and ecological civilization based on local industrial development and natural endowment and aiming at improving the quality of air, water and soil. In 2022, according to the monitoring results from surface water body sections under national and provincial monitoring programs of the Yangtze River, the Yellow River and other large rivers in Qinghai Province, the proportion of Grade II water quality topped 76%, meaning "a river of clear water will flow downstream" and people can enjoy the scenery of "green water around the city". In other words, the achievements of ecological civilization construction have been reflected in a more beautiful ecological environment and further improving happiness, sense of achievement and pride of the people.

The CPC and the Chinese government emphasize that the ecological environment is one of the "most equitable public goods", attaching importance to the results of environmental governance as well as the primary responsibility for governance. This primary responsibility of governments means that members of the CPC and officials in various Chinese governments are required to firmly abandon the narrow view of performance and people's livelihood that focuses on GDP only and to establish a new view of performance, requiring the CPC and the Chinese government to take a good ecological environment as a basic public service, to implement green and scientific decision-making, to effectively update the concept of governance with the new development philosophy, to continuously improve the quality of the ecological environment as an important way to fulfil the responsibility of ecological civilization construction, to strengthen public governance, to pursue the most universally beneficial welfare, and to meet the growing needs for ecological products of the people in all aspects.

3.3 Better Demonstrating People-Centered Development Philosophy Through Boosting High-Quality Development

At this stage, the principal contradiction facing Chinese society has been transformed into the contradiction between people's ever-growing needs for a better life and unbalanced and insufficient development, while the contradictions and problems related to development are mainly concerned with the development quality. During the 14th Five-Year Plan period, China will achieve new industrialization, IT application, urbanization and agricultural modernization and complete building socialist modernization in all aspects in a country with a population of 1.4 billion. High-quality development means that development focuses more on "quality" rather than "quantity", and China will put people first and fully carry out the people-centered development philosophy by consciously boosting transformation toward green and high-quality development so that the people will realize and feel it helpful to achieve more equitable, more sustainable and safer development and build China into a beautiful country by improving the ecological environment through promoting high-quality development.

4 Principles of Ecological Civilization

The ecological environment is a major political issue related to national governance and mission and a key social issue concerning people's livelihood. The CPC and the Chinese government have always attached great importance to ecological and environmental protection and have established resource conservation and environmental protection as a basic state policy and sustainable development as a national strategy.

4.1 Harmonious Coexistence Between Humans and Nature

Humans and nature form a biotic community with a shared future, humans must live on nature, and the lifeblood of human survival and development is in the ecosystem. The history of nature and humans are essentially closely related: a civilization may thrive if its natural surroundings thrive and will suffer if its natural surroundings suffer. *Dialectics of Nature*, written by Friedrich Engels, shows that the people who, in Mesopotamia, Greece, Asia Minor, and elsewhere, destroyed the forests to obtain cultivable land, never dreamed that they were laying the basis for the present devastated condition of these countries by removing along with the forests the collecting centers and reservoirs of moisture. The direct consequence of human destruction of ecology is the annihilation of civilizations and the end of people.

To construct ecological civilization, China insists on the harmonious coexistence of humans and nature, which is a high summary of history and objectivity. Human history has profoundly and dialectically revealed that it has to coordinate economic and social development with ecological construction since both human history that focuses on people only and cuts off natural history and natural history that focuses on nature and cuts off human history are one-sided. While upholding the idea of a shared future for humans and nature, China should not only adhere to the principles of respecting, accommodating to and protecting nature but also carry out the "people-centered development philosophy" to integrate the "naturalism of humans" and the "humanism of nature".

4.2 Invaluable Assets of Lucid Waters and Lush Mountains

The idea that lucid waters and lush mountains are invaluable assets explains the relationship between economic development and ecological environmental protection, reveals the truth that protecting and improving the ecological environment equals the protection and development of productive forces, and indicates a new path to achieve synergy between development and protection. Lucid waters and lush mountains are both natural and ecological wealth and social and economic wealth. Protecting the ecological environment is to conserve the natural value and appreciate the value of natural capital and to protect the potential and momentum for economic and social development, and it will ensure that lucid waters and lush mountains continue to contribute to ecological civilization construction and economic-social development.

Caohu Town, in Tumushuke City of Akesu Prefecture, Xinjiang Uygur Autonomous Region, was an important communication hub of the ancient Silk Road and a bridgehead of the China-Pakistan Economic Corridor. Through various construction projects of "Ecological Caohu", "Beautiful Caohu", "Humanistic Caohu", "Dynamic Caohu" "Happy Caohu", and "Strong Caohu", Caohu Town vigorously promoted the development of green industries, including environmentally friendly market gardening with high ecological benefits, namely, support for the cultivation of strawberries, nectarines, seedless watermelon and other anti-seasonal fruits and vegetables, and boosted local economic development through ecological construction. Therefore, the town has been entitled as a National Civilized Town and a National Characteristic Town and selected as a "lucid waters and lush mountains are invaluable assets" practice and innovation base by the Ministry of Ecology and Environment.

Caohu Town

Source Convergence Media Center of the Third Division of Xinjiang Production and Construction Corps (XPCC), "Green Development Improves the People's Well-beings and Happiness, the Practice of 'Lucid Waters And Lush Mountains Are Invaluable Assets' in Caohu Town by 41st Regiment, Third Division of XPCC", http://www.btzx.com.cn/web/2022/2/14/ARTI1644815490546294.html

Now, the concept of "lucid waters and lush mountains are invaluable assets" is becoming the philosophical idea and practical paradigm for China to coordinate development and protection, ecology and economy, economic-social development and environmental protection. In the past China developed its economy focusing on GDP only, did not establish the concept of natural values and natural capital, and considered development and protection to be opposite, believing that environmental protection would prevent economic development. Therefore, China used to pollute first and clean up later, clean up and pollute at the same time, or pollute without cleaning up, which cannot last for now. Through the construction of ecological civilization, all of Chinese society has been well aware that good environmental quality indicates a high-quality economy while promoting and enhancing social modernization. The CPC, the whole country and all sectors of society highly agree that by upholding the idea that lucid waters and lush mountains are invaluable assets, China can explore a new road to high quality and sustainable green development integrating economic development and environmental protection.

4.3 Good Eco-Environment is the Most Universally Beneficial Welfare

The environment is crucial to the wellbeing of the people since blue skies and verdant surroundings allow people to feel the beauty and joy of life. Economic development is for the wellbeing of the people, as is environmental protection.

The Xinjiang Uygur Autonomous Region is vast, and the climate there is pleasant. Therefore, the atmospheric environmental problem due to the industrialization process is particularly prominent and has become a social concern. Therefore, Xinjiang carried out comprehensive control of three major pollution sources of industry, coal-burning plants and automobiles, and steadily advanced the promotion of clean energy and emission-reducing technical upgrade projects. In 2021, Xinjiang invested USD 207.6 million as special funds to support clean energy heating and to help enterprises carry out projects such as coal burning and industrial pollution control, boiler and industrial kiln all-round improvement, and volatile organic compound (VOC) treatment, while boosting 295,400 households to replace coal with electricity and further expanding the replacement of coal with electricity projects to five prefectures in southern Xinjiang and these conditional counties and cities in northern Xinjiang and eastern Xinjiang. Under the guidance of ecological civilization ideology, blue sky and white clouds have once again become the most familiar scenery of residents in Xinjiang.

Insisting on ecological construction for and benefiting the people, China makes greater efforts to improve the insufficient supply of ecological products in the field of people's livelihood and to boost the three major action plans to control air, water and soil pollution, including controlling air pollution in key industries and key regions, preventing and controlling particulate pollution, treating watershed and regional water pollution, and addressing heavy metal pollution and soil pollution. These measures significantly strengthen ecological and environmental governance in China, improve the eco-environment, and enrich ecological products, ensuring that people enjoy a higher quality eco-environment of lucid water and lush mountains, as well as improving welfare and the sense of contentment.

4.4 Upholding a Biotic Community of Mountains, Rivers, Forests, Fields, Lakes, Grasslands and Deserts

The biotic community is the material basis for the survival and development of humans. For a period of time, due to the lack of a systematic, holistic and long-term view in promoting ecological environmental protection and pollution management and not fully recognizing the value of the biotic community and ecosystem of mountains, rivers, forests, fields, lakes, grasslands and deserts, China carried out segmented pollution prevention and control practices, from the surface to the underground, from the banks and shores to the rivers and lakes, from the land to the sea, from the cities to the countryside, and from carbon monoxide to carbon dioxide, resulting in confusing division of responsibilities between different ministries, and treatment of superficial symptoms through stopgap measures with government departments each looking out for their own immediate problems. Therefore, although China made huge investments in human, material and financial resources, there were no obvious results, but new pollution problems.

Since the 18th CPC National Congress, China has emphasized the need to explore new ways of pollution governance from the perspective of system engineering and a holistic approach and to make plans that take all factors into consideration and simultaneously implement multiple comprehensive measures to ensure that efforts to build an ecological civilization permeate all fields, regions, and processes.

In September 2018, General Secretary of the CPC Central Committee, Xi Jinping, issued a prescription for the protection of black soil in northeast China: "Adhere to the combination of utilization and nourishment, while apply comprehensive measures, to ensure that black soil would not decrease or degrade." Farmers in Lishu County, Siping City, Jilin Province, now are used to returning straw of crops that are just harvested to the cultivated land, to maintain moisture and cultivate humus, as they have explored a replicable ecological restoration program. Around the village of Gaoxigou in Mizhi County, Yulin City, Shaanxi Province, these high slopes of loess soil have been covered by pine, cypress and acacia trees as well as fruit trees such as apple, apricot and jujube, bringing the forest and grass coverage rate to over 70%. Through decades of hard work, villagers solved the ecological issue of loessland, which is barren and prone to loss of water and soil. Ganzhou in Jiangxi Province controlled the soil erosion of more than 3,400 km^2 during the 13th Five-Year Plan period through water and soil conservation and afforestation, turning the former "red desert" into a green forest.

Gaoxigou Village

Source Yulin Media Center, "Visual Shaanxi: Gaoxigou Village in Mizhi County, Yulin City: A Village with Typical Landscapes in Southern China of Lucid Waters and Lush Mountains Appears in Western China", http://ishare.ifeng.com/c/s/v002sgbBeG-_hy1QL4-_s3Aj9-9f1OfrCIHHo9by6v4CVut9A

For the integrated conservation and restoration of mountains, rivers, forests, farmland, lakes, grasslands, deserts, glaciers and snow mountains, China has carried out a series of fundamental, groundbreaking and long-term projects, with unprecedented determination and efforts, realizing great achievements. In particular, China has vigorously promoted and implemented major strategies such as the development of the Yangtze River Economic Belt and the ecological protection and high-quality development of the Yellow River Basin to boost the comprehensive governance of watershed ecosystems, setting up a model of insisting on systematic management.

4.5 Protecting the Eco-Environment with the Strictest Possible Regulations and Laws

The protection of the ecological environment must rely on legal and regulatory backing. China has introduced a series of reform measures and related institutions to firmly establish institutional rigidity and authority in the construction of ecological civilization, leaving no room for choice, flexibility, or compromise and never letting institutions and regulations become "toothless tigers".

On the Earth Day in 2021, the Huqiu Court of Suzhou in Jiangsu Province announced a number of typical cases concerning environmental resource protection in hopes of further playing the role of judicial adjudication in protecting environmental resources, educating the whole society with ecological protection ideas, and protecting the ecological environment with regulations and laws, the most universally benefiting welfare. In recent years, under the guidance of Xi Jinping Thought on eco- civilization and following the requirements of the Grand Canal Basin ecological protection and high-quality development strategy, the Huqiu Court made contributions to the battle of pollution prevention and control by providing judicial assistance, handling various cases involving environmental resources according to law, and comprehensively promoting the overall improvement of the ecological environment and the legal and reasonable use of natural resources.[2]

China protects the ecological environment with the strictest regulations and laws and has constructed a complete and effective institutional framework, which is composed of a system of property rights for natural resource assets, a system for the development and protection of territorial space, a spatial planning system, a system for regulating total consumption and comprehensive conservation of resources, a system for payment-based resource consumption and compensating conservation and protection efforts, the environmental governance system, the market system for environmental governance and ecological preservation, and an ecological conservation performance evaluation and accountability system, designed to promote ecological progress with clearly defined property rights, diversified participation, and equal focus on incentives and restraints. Meanwhile, China will modernize the governance system and capacity for governance in the field of ecological progress by resolutely correcting institutional drawbacks and striving to carry out the reform in a systemic, more holistic, and better coordinated way.

4.6 *Joining Efforts in Building a Global Ecological Civilization*

The construction of ecological civilization concerns the future of humanity, and building a green living environment is humanity's common aspiration. The whole world must work together to protect the environment and respond to climate change. No country can ignore these issues, and no country will fare well by going it alone. However, in recent years, economic globalization has encountered challenges, as more unilateralist and protectionist policies are issued. China advocates the construction of ecological civilization and contributes to the speeding-up cultivation of global ecosystems that respect nature and focus on green development and to a cleaner and more beautiful world.

---

[2] *Good Eco-Environment Is the Most Universally Beneficial Welfare, Huqiu People's Court Re leased Some Typical Cases of Environmental and Recourse Protection*, The People's Court in Huqiu District, Suzhou City, https://www.thepaper.cn/newsDetail_forward_12344477.

First, China upholds the vision of a community with a shared future, in which all countries and regions are interdependent. As China proposed, all countries and regions around the world are closely interconnected with a shared future for mankind. No country can gain from others' difficulties or maintain stability by taking advantage of others' troubles. To pursue a beggar-thy-neighbor policy or just watch from a safe distance when others are in danger will eventually land one in the same trouble faced by others. Therefore, it is necessary to embrace the vision of a community with a shared future in which everyone is bound together, reject attempts to build blocs to keep others out, oppose a zero-sum approach, see each other as members of the same big family, and pursue win–win cooperation. Countries must cooperate to address global environmental issues such as climate change, marine pollution and biological conservation and to achieve the UN Sustainable Development Goals 2030.

Second, China abides by the principle of common but differential responsibility and provides more assistance to developing countries, especially small island countries. China has repeatedly stressed that addressing climate change is an inherent requirement for China's sustainable development and an international obligation for a responsible major country, which is not a request placed upon China but a choice China made freely. China insists on justice and equity, benefit sharing, and caring about the concerns of developing countries in terms of finance, technology and capacity building. China abides by the principle of common but differential responsibility and has provided targeted assistance in terms of finance and technology to developing countries, especially small island countries.

China plans to realize the goals of peaking carbon dioxide emissions by 2030 and achieving carbon neutrality by 2060, marking both the construction of ecological civilization and economic-social development in China, which have entered a new era in which China sets carbon emission reduction as the key strategic target from a global perspective. China will certainly coordinate domestic and international efforts to achieve synergy and efficiency in reducing pollution and carbon emissions as the key to promoting comprehensive green transformation of economic and social development to make China and the world more beautiful.

Chapter 2
A New Road to Modernization: Modernization Ensures Harmoniousness Coexistence Between Humans and Nature

Meng Li, Guoqing Li, and Shouxian Zhu

Since the 17th century, it has taken more than three hundred years for developed countries in the West to modernize approximately one billion people and map out the Western path of modernization. The differences in historical conditions, geographical environment, national culture, and development stages determine the diversity of development paths and roads to modernization. Marx once suggested that Eastern societies could cross the capitalist "Caudine Valley"[1] and open up another road to modernization. Since the founding of the People's Republic of China, the CPC and the Chinese government, based on the national conditions and the political tradition, economic structure, cultural genes and historical context of the Chinese nation, have come out with a new road to modernization unique to Chinese characteristics and different from that of the West and have created a new form of human civilization, ecological civilization, in the course of promoting modernization that ensures harmonious coexistence between humans and nature, turning Marx's theory into reality and breaking the absolute monopoly of the Western modernization model fundamentally by developing a new theoretical form of modernization and providing the world with a new option through setting a successful example.

[1] A metaphor for disastrous historical experiences, and it can be used to describe the great difficulties and challenges people face when they seek development.

M. Li (✉) · S. Zhu
Research Institute for Eco-Civilization, Chinese Academy of Social Sciences (CASS), Beijing, China

G. Li
School of Ethnology and Sociology, Minzu University of China, Beijing, China

K. Yang (ed.), *The Construction of Ecological Civilization in China*, China Insights,
https://doi.org/10.1007/978-981-99-6515-1_2

1 A Modernization that Ensures Harmonious Coexistence Between Humans and Nature

Modernization is the process toward and the state of a modern society transformed from a traditional one, symbolizing the development level of human civilization and the common cause of all humans. China's modernization is an important part of the modernization of the world. The CPC has continuously deepened its understanding of the connotation of and the road to modernization and gradually found a path to modernization with Chinese characteristics, a modernization in which humans and nature can develop in harmony, uniting and leading all Chinese people.

1.1 The Idea of Modernization that Ensures Harmonious Coexistence Between Humans and Nature

The idea of new modernization that ensures harmonious coexistence between humans and nature emerges during a special historical period and has rich time significance. The modernization process first started in the 17th century in the West, driven by industrialization, and boosted the rapid transformation of traditional agricultural societies into modern industrial societies.[2] Rapid industrialization, driven by technological advancements, oriented by economic development, and supported by massive exploitation of resources, was dedicated to the transformation of nature and the pursuit of wealth accumulation, since the material wealth it produced over 300 years far exceeded the sum of material assets created by humans ever before. However, entering the 1960s, this modernization oriented by industrial civilization caused tension between humans and nature and serious environmental problems, forcing humans to reflect on the boundaries for resources and development and the relationship between humans and nature. The Anthropocene Working Group[3] (AWG) under International Commission on Stratigraphy (ICS) noted that humans have dominated the symbiotic relationship between humans and nature, the biotic community between humans and nature, and the changes in global ecosystems and that the Earth has moved from the "Holocene" to the human-dominated "Anthropocene". In the Anthropocene, humans are strong enough to regulate the symbiotic relationship between humans and nature and to determine the shared future of the human-nature community and the ecosystems on Earth. In terms of practical logic, the Earth has been in a state of emergency. As stated in the *World Scientist's Warning to Humanity*, "Humanity and nature are in a process of conflict with each other. Human activity

---

[2] Such profound changes were all-round, as the traditional economy and society were modernized, and traditional agricultural civilization developed toward modern industrial civilization. The impact is transmitted globally. In the nineteenth and twentieth centuries, there were waves of modernization construction in South America, Asia and Africa.

[3] The AWG was formed in 2016 by the International Commission on Stratigraphy, and it includes dozens of geoscientists.

has caused serious and irreparable damage to the main environmental resources. If left unchecked, many of our actions will put the human society we expect for in the future and the world of plants and animals in jeopardy, and will dramatically alter this home for life".[4] How to regulate the relationship between humans and nature with human strength and whether to follow the old path of industrial civilization or seek a new development road to civilization have become global and fundamental issues facing human society in recent years.

Like many other countries in the world, China also experienced a historical process of advancing into nature, transforming nature and conquering nature along with industrialization, resulting in an increasingly prominent contradiction between humans and nature while rapidly forming the material basis for modern development. In addition, resources and the environment have become bottlenecks that seriously restrict development. Faced with the global challenges in the Anthropocene and the increasingly sharpening domestic contradictions, including the worsening relationship between humans and nature, the CPC, while focusing on solving the social contradiction between the ever-growing needs of the people and unbalanced and insufficient development, has made it clear that economic development must not be at the expense of the ecological environment and that it should resolutely discard these development models and practices that damage or even destroy the ecological environment. Modernization with Chinese characteristics should not follow the old path of modernization by industrialized countries. China has taken the lead in building a new form of ecological civilization and exploring a new path to modernization in which humans and nature can coexist in harmony.

In 2012, it was at the 18th CPC National Congress that the construction of ecological civilization was incorporated into the "five-sphere integrated plan" for building socialism with Chinese characteristics, proposing to "vigorously promote the construction of ecological civilization and the harmony between humans and nature". At the Third Plenary Session of the 18th CPC Central Committee, it was agreed to incorporate the "reform for promoting ecological progress" into the objectives of comprehensive deepening reform and proposed to achieve a new approach to modernization that is characterized by the harmonious coexistence of humans and nature. In 2018, the *Amendment to the Constitution of the People's Republic of China* that was adopted at the National People's Congress (NPC) modified the description for the basic task of China as China would "promote coordinated material, political, cultural-ethical, social and ecological advancement, in order to build China into a great modern socialist country that is prosperous, strong, democratic, culturally advanced, harmonious and beautiful, and realize the great rejuvenation of the Chinese nation" and deleted the old one that China would "promote the coordinated development of the material, political, and spiritual civilizations to turn China into a powerful and prosperous socialist country with a high level of culture and democracy". Thus,

[4] "Scientist Statement: World Scientists' Warning to Humanity" (1992) (PDF document), issued on July 16, 1992 and updated on October 29, 2002.

the CPC has systematically defined the position of the new form of ecological civilization in China's overall plan for advancing socialism with Chinese characteristics, the coordinated development of civilizations and the nature of a modern state.

In 2017, the report of the 19th CPC National Congress proposed to "uphold the harmonious coexistence between humans and nature" and established it as one of the basic strategies for adhering to and developing socialism with Chinese characteristics. In 2018, at the National Conference on Environmental protection, "upholding the harmonious coexistence between humans and nature", became a basic principle to advance ecological civilization construction, as "the modernization that we pursue is one characterized by harmonious coexistence between humans and nature. In addition to creating more material and cultural wealth to meet people's ever-increasing needs for a better life, we need also to provide more quality ecological goods to meet people's ever-growing demands for a beautiful environment". In 2020, at the Fifth Plenary Session of the 19th CPC Central Committee, the CPC Central Committee once again clarified that China would "construct an ecological civilization, promote the comprehensive green transformation of economic and social development, and build a modernization in which people and nature live in harmony". To date, the CPC has formally established the position of modernization featuring harmonious coexistence between humans and nature in the overall construction of modernization with Chinese characteristics.

1.2 A New Road to Modernization that Ensures Harmonious Coexistence Between Humans and Nature

The modernization that ensures harmony between humans and nature is one that is holistically planned and follows the basic principle of such harmonious coexistence. It marks the new journey of Chinese modernization and creates a new form of civilization, ecological civilization. Unlike the modernization of Western industrial civilizations, the modernization that ensures harmony between humans and nature is unique and advanced in terms of the value idea, development mode and vision.

In terms of value idea, the traditional theories of modernization attach more importance to economic value and ignore the ecological value or regard ecological value as the value of an economic factor for the contribution to industrial society of environmental resources, thus making it difficult to fully realize ecological value and human value and turning relations between humans and nature to be antagonistic or alienated. In contrast, modernization, which ensures harmonious coexistence between humans and nature, emphasizes the value of ecology itself, which includes not only the ecological service value of natural resources such as forests and rivers but also important economic values from the ecological agricultural economy, tourism economy and so on. Humans can realize their value through rationally utilizing and transforming nature. Modernization to realize the harmonious coexistence between humans and nature is based on the systematic and holistic consciousness that nature, society and individuals can develop as an organic whole through organic interaction.

In terms of the development mode, although it has greatly improved the productivity and correspondingly the living standards of the residents, the traditional industrialization process has also greatly intensified the grabbing and plundering of natural resources and brought about serious damage to the ecological environment, for extensively expanding the mass production mode relying on resource capital as a production factor. Modernization that ensures harmonious coexistence between humans and nature requires a comprehensive green transformation of the production and living modes, gives priority to ecology and green development, placing "green" before "development", and carries out effective green transformation through the production and consumption chains to promote the dialectical unity of economic development and ecological environmental protection.

In terms of vision, traditional modernization based on the industrial revolution aims at maximizing profits, pursuing the conquest of nature and the law of the jungle, resulting in the continuous accumulation of man-made material wealth. At the same time, the huge wealth created during industrial development benefited only a small group of people at the top, ignoring the majority of people at the bottom. The modernization that ensures harmony between humans and nature, on the other hand, takes a more integrated approach, meaning it pursues not only the achievements of a higher quality modernization and ecology but also the benefits to the people with these achievements to effectively enhance the people's satisfaction, happiness, and sense of achievement. The modernization that ensures harmony between humans and nature targets at common prosperity that is both material and spiritual, and for all the people, and calls for win–win cooperation and common development of all countries, rather than unilateralism and protectionism, to build a community with a shared future for mankind and a beautiful Earth.

In *Beyond Civilization*, Paul Bohannan said, "It is obvious that we stand on the threshold of a postcivilization, and that the society and culture we will build would be a world that has never been seen before, after we solve the problems we face today. It may be somehow more civilized than what we already have, but it will certainly not be the civilization we already know."[5] China advocates modernization that ensures harmonious coexistence between humans and nature, which will bring reform to the development paradigm and will certainly drive a reshaping of social civilization.

The Chinese nation has always been loving, respecting and accommodating to nature. In the history of Chinese civilization, which lasted for more than 5,000 years, the ancestors of the Chinese nation had realized the importance of the ecological environment long before and developed wise ecological philosophy. Confucius once said: "子钓而不纲, 弋不射宿," meaning using a pole rather than a large net to catch fish and shooting arrows at birds in the air not those in the nests. Xuncius said: "草木荣华滋硕之时则斧斤不入山林, 不夭其生, 不绝其长也; 鼋鼍、鱼鳖、鳅鳝孕别之时, 罔罟、毒药不入泽, 不夭其生, 不绝其长也 (It means when the grass and

[5] Paul Bohannan, "The State of the Species", *Beyond Civilization*, Natural History-Special Supplement 80, 1971.

trees are blooming and fruiting, it should not allow to carry axes into the mountains and forests, to prevent disturbing the birth and growth of plants, and when the turtles, crocodiles, fish, and other aquatic life are pregnant and breeding, it should not allow to throw nets or poisons into waters, to prevent disturbing the birth and growth of animals)." It also states in Lu Shi Chun Qiu (Mister Lv's Spring and Autumn Annals) that "竭泽而渔, 岂不获得?而明年无鱼; 焚薮而田, 岂不获得?而明年无兽 (How couldn't it be fruitful if fishing by draining the pond? However, there will be no place to fish next year. How couldn't it be fruitful if hunting by burning down the forest? However, next year there will be no place to hunt)." All these philosophies bore the same idea that humans should make use of nature in a timely and moderate manner, which is of great practical importance. Likewise, the philosophical concepts of "man is an integral part of nature" and "Dao follows the laws of nature" and the family-regulating motto of "一粥一饭, 当思来处不易; 半丝半缕, 恒念物力维艰 (save food by keeping in mind that every meal and grain of rice is the fruit of hard work, and care for clothes by reminding ourselves that every thread is the product of toil)", these simple and wise views of nature are still profoundly meaningful and inspiring today.

China's path to modernization, in which human beings and nature live in harmony, not only carries on the genes of Chinese civilization but also focuses on simultaneously promoting the construction of material and ecological civilization, essentially embodying the ecological wisdom of the Chinese nation. To implement the new development philosophy completely, accurately and comprehensively, China should adhere to the basic state policy of resource conservation and environmental protection and uphold the principles of prioritizing resource conservation and environmental protection and letting nature restore itself to develop spatial layouts, industrial structures, and ways of work and life that help conserve resources and protect the environment, striving for socialist modernization.

2 A Road Prioritizing Ecology and Green Development

To construct modernization that ensures harmonious coexistence between humans and nature, China must be determined to "put ecology first and remain committed to green development", following the objective laws of nature and emphasizing the subjective initiative of humans. Prioritizing ecology requires respecting and caring for nature, and green development indicates that it should give full play to human initiative to seek the development of human society and the scientific way of development. As Chinese leaders pointed out, "practices have proved that economic development cannot be achieved at the expense of ecology, ecology itself is the economy, and by protecting the environment we are protecting productive forces, and by improving the environment we are developing productive forces", and "a civilization may thrive if its natural surroundings thrive, and will suffer if its natural surroundings suffer". Only by respecting, accommodating to and protecting nature and ensuring harmonious coexistence between humans and nature can humans find a new road to modernization where economic development and ecological civilization complement each other and bring out the best in each other.

2.1 *Constructing a High-Quality Eco-Environment*

To construct a modernization in which humans and nature live in harmony, first of all, it needs to correctly handle, the relationship between human production and living activities and ecological environmental protection needs to be correctly handled. Unlike the classic Western modernization transforming a traditional agricultural society into an industrial society, modernization with Chinese characteristics intends to follow a path to civilization that enables sustainable development, prosperity and good ecology, rejecting the "modernization" model of disregarding, dominating and destroying nature and the old path of traditional modernization of "polluting first and fixing it later".

The CPC Central Committee has made it clear that ecological environmental protection is a cause that will benefit generations. Upholding the basic view of people's livelihood that a good ecological environment is the most universally benefiting welfare, the CPC has led the nation to implement three major action plans to improve air, water, and soil quality, set ecological conservation red lines, optimize the industrial layout and structure, put scattered, noncompliant, and polluting under unprecedented scrutiny, and coordinate these projects for ecological protection, energy saving and emission reduction, environmental improvement, and building beautiful homeland, while actively carrying out ecological demonstration projects, strengthening the construction of nature reserves, intensifying the protection of biodiversity, and continuously implementing key projects such as returning pasture to grass, protecting wetlands, and building protective forest systems. As a result, China's ecological environment has improved significantly, reflected in an increasing number of blue skies, clearer rivers and better ecology. By the end of 2021, the ratio of days with good air quality in prefecture-level cities and above reached 87.5%, up 0.5 percentage points year-on-year; the concentration of fine particulate matter (PM2.5) was down 9.1% year-on-year, to 30 micrograms per cubic meter; surface water body sections that met Grade I–III water quality standards was 84.9%, up 1.5% points year-on-year; and the reduction target for carbon dioxide emissions per unit of GDP is expected to meet the sequential progress requirements of the 14th Five-Year Plan.[6] According to the Ministry of Ecology and Environment of China, eight obligatory

[6] "National Eco-environmental Quality Improved Significantly in 2021", *Xinhua News Agency*, http://www.gov.cn/xinwen/2022-04/18/content_5685888.htm.

ecological and environmental targets set in the 2021 National Economic and Social Development Plan were successfully achieved, marking a good start for the 14th Five-Year Plan period in ecological environmental protection.

Frederic Baldan, CEO of the Belgian consulting company CEBiz who visited China many times, expressed his intuitive feelings about improving the ecological environment in China: "There are more and more days with good air quality in Chinese cities, and the air is getting cleaner and cleaner." He said that China has given the strongest policy support to energy saving and clean energy development and carried out scientific and pragmatic integrated management of mountains, rivers, forests, fields, lakes, grasses and deserts, making remarkable achievements in the construction of ecological civilization. Masamichi Mitsuma, professor emeritus of Reitaku University in Japan, has been watching the development and changes in China for more than 40 years. In his view, since the 18th CPC National Congress, the Chinese government has attached great importance to and scientifically addressed environmental issues such as air, water and soil pollution and has taken a series of effective measures that have improved the environment. During an interview on the eve of the Fifth United Nations Environment. Assembly on February 22nd, 2021, executive director of United Nations Environment Programme (UNEP), Inger Andersen, highly appreciated the great progress made in the construction of ecological civilization in China. She said, the core target of China's ecological civilization construction is to achieve harmony between humans and nature, which is also the vision of all countries and regions around the world.

Yu village in the 1980s

Yu Village in 2018

Source thepaper.cn, "The Experience of Zhejiang in Practising the Idea of 'Lucid Waters and Lush Mountains': Making Good Use of the Beautiful Eco- Environment Can Realize Remarkable Spill-over Effects", https://ishare.ifeng.com/c/s/7z12umH9udW. (the picture above)Zhengguan News, "UNWTO Included Two Villages in China in the First Batch of Best Tourism Villages", https://baijiahao.baidu.com/s?id=1718162308866107915&wfr=spider&for=pc. (the picture below)

China has taken a new path to modernization by coordinating economic development and environmental protection, improving people's sense of gain, happiness and security from the ecological environment in the modernization process. The two pictures show the fruits of beautiful rural construction in China, namely, the transformation of Yu Village in Tianhuangping Town, Anji County, Huzhou City, Zhejiang Province.

2.2 *Taking Efforts to Advance All-Round Green Transformation*

Green transformation means adopting a green development model that enables a mutual promoting relationship between economic development and resource savings, emission reduction and environmental improvement, replacing the old model where economic development relies on high consumption, high emissions and damage to the environment. The transformation is reflected in many aspects, including the concepts and ideas, value orientation, spatial layout, production mode and lifestyle. Green development is not a repair of the traditional industrialization model but an all-round and systematic green transformation in the new period of modernization.

In 2015, the Fifth Plenary Session of the 18th CPC Central Committee established the concept of green development as an important development concept to follow for a long time to come, making it clear that green development is an important guideline in promoting modernization. Accordingly, the CPC Central Committee has made a series of strategic deployments, emphasizing that only by upholding the concept of green development and pursuing a model of sustainable development featuring increased production, higher living standards and healthy ecosystems can China effectively overcome the hard constraints of resources and the environment on building a moderately prosperous society in all aspects while laying a solid ecological foundation and providing a powerful engine for building a strong modern socialist country. These strategic deployments mainly include measures in the following areas.

Industrial structural reform and green industrial development come first. Optimizing the industrial structure and improving the industrial chain is an important way to promote green economic transformation. While speeding up industrial modernization, China is vigorously boosting structural reform to resolutely eliminate obsolete excess capacity in coal, iron and steel, cement, plate glass, electrolytic aluminum and other industries, update technology and equipment, reduce energy consumption and emissions, and accelerate the green transformation and upgrade of traditional industries. Meanwhile, China has made great efforts to cultivate strategic emerging industries, including energy conservation and environmental protection, new energy, new-generation information technology, biology, new materials, new energy vehicles and so on, develop green services, introduce energy performance and water conservation contracting systems, and construct a green industrial system. In recent years, China has actively advanced the development of intelligent manufacturing, developed "Internet +" manufacturing models in great numbers, and introduced the industrial Internet widely into oil, petrochemical, steel, home appliances, clothing, machinery, energy and other industries, providing a strong impetus for the green transformation of the manufacturing industry. The *Plan for National Informationization during the 14th Five-Year Plan Period* further proposed to "further promote the construction of green and smart ecological civilization, and the synergistic digital and green development", and "boost green development with digital technology, and drive digital transformation with green development". The Internet, big data, artificial intelligence, fifth-generation mobile communications (5G) and other emerging technologies are accelerating deep integration with green and low-carbon industries, providing new momentum for high-quality economic and social development in China.

Second, China boosts energy structural adjustment and new energy system construction. Energy is an important material basis for the national economy (the energy industry itself is also an important part of the national economy), a key support for national security and development as well as the people's well-being, and an indispensable foundation and guarantee for the modernization process. Historically, the ways to acquire energy, the conversion and utilization efficiency, and per capita consumption were important demonstrations of a country's production technology, standard of living and international competitiveness, as well as an important symbol of human civilization. Due to China's coal-dominating resource endowment, energy structural adjustment and modern energy system construction will be of great importance in green transformation. On the consumption side of energy, it has become a social consensus to control the consumption of fossil energy. General Secretary Xi Jinping proposed to "tighten the control of total energy consumption, effectively carry out the principle of prioritizing energy conservation, enforce energy conservation in all areas of social and economic development, resolve to adjust the industrial structure, emphasize energy conservation in the process of urbanization, advocate hard work and thrift, and move faster towards an energy-saving society."[7] To carry out the deployment of the central government, relevant ministries and commissions successively issued the *National Energy Development Strategy Action Plan (2014–2020)* and the *Energy Security Strategy of China*, the *Strategy of Energy Production and Consumption Revolution (2016–2030)*, to promote the transformation of energy production and consumption, while adopting the *Air Pollution Prevention and Control Action Plan*, the *Work Plan for Strengthening Air Pollution Prevention and Control in the Energy Sector*, and implementing "new towns, new energy and new life" action plans to promote the urban and rural energy structural adjustment and the improvement of energy efficiency. On the supply side, China vigorously promotes the clean and efficient use of coal, and focuses on the development of noncoal energy, especially the renewable energy industry. In the last ten years or so, the average cost per kilowatt of onshore wind and photovoltaic power generation projects in China has dropped by approximately 30% and 75%, respectively, marking continuously improving industrial competitiveness, and the newly installed and cumulative capacity of wind and photovoltaic power has led in the world for years. In 2021, the newly installed capacity of renewable energy in China totaled 134 million kilowatts, accounting for 76.1% of the total newly installed capacity of power generation, and electric energy production reached 2.48 trillion kilowatt hours, accounting for 29.8% of the total electricity consumption. The installed capacity of renewable energy exceeded one billion kilowatts, both the installed capacity of wind and photovoltaic power topped 300 million kilowatts and that of offshore wind power took the lead around the world, while the installed capacity of all these powers mentioned above accounted for more than one-third of the world total. A new green energy system is taking shape in China.

Third, China advocates green transformation in consumption and lifestyle. In recent years, the Chinese government has put green consumption and lifestyle in a more prominent strategic position and has taken a series of policies and measures to promote high-efficiency lighting and other energy-saving products, popularize water-saving products including faucets, toilets and washing machines, and new energy vehicles by accelerating the construction of electric vehicle charging infrastructure. Between 2012 and 2016, the cumulative amount of government procurement for energy-saving (water-saving) products in China reached USD 103.25 billion. The number of green consumers on Ali's retail platform grew 14 times between 2012 and 2015. It was estimated that the energy-efficient air conditioners, refrigerators, washing machines, flat-panel TVs and water heaters sold in 2017 could save electricity consumption by approximately 10 billion kWh a year, equivalent to

[7] Xi Jinping, "Vigorously Promote Revolution in Energy Production and Consumption", *People's Daily*, June 14, 2014.

an emission reduction of 6.5 million tons of carbon dioxide, 14,000 tons of sulfur dioxide, 14,000 tons of nitrogen oxides and 11,000 tons of particulates. Various green projects were implemented in various regions, including action plans for green families, schools, communities, shopping malls, restaurants and so on, advocating green living by saving water and electricity and setting reasonable indoor temperatures for air conditioning in summer and heaters in winter, vigorously developing public transportation, encouraging green travel such as bicycles and walking, and establishing a household waste classification system, to urge residents to participate widely in waste sorting and recycling. The green lifestyle stimulates the supply of green products and services, which in turn further promotes the green transformation of production.[8]

Fourth, China is strengthening green science and technology and innovation capabilities. Modernization that ensures harmonious coexistence between humans and nature pursues an upgrade in the relationship between humans and nature, and science and technology innovation will play a key role in promoting green transformation. Since the reform and opening-up, China has implemented the strategy of invigorating the country through science and education, and investment in science and technology has increased significantly. In 2014, China surpassed Japan and the EU and became the world's second largest economy in terms of R&D investment, contributing nearly a quarter of global total R&D expenditure. In 2018, the total R&D expenditure in China was nearly USD 246.8 billion, accounting for 2.19% of GDP, exceeding the average (2.1%) in the 15 EU countries. During the same period, R&D investment in green technologies in China also increased dramatically. Between 1990 and 2014, the number of environment-related patents in China increased 60-fold compared to a threefold rise in OECD countries, and the number of patent applications for "green" technologies in China rose at a higher rate than that for all technologies over the past decade, especially since 2005.[9] Meanwhile, the Chinese government has supported the development of green technologies through institutional construction to provide a favorable environment, encouraged cooperation among innovation bodies and fully stimulated their initiative to tap the potential. In addition, intermediary service institutions, innovation parks, green technology R&D centers and other platforms are gathering innovation resources, and the construction of new-generation innovation infrastructure centered on the Internet has progressed. China continues to strengthen its capacity to convert advancements in science and technology into actual productive forces. In recent years, China alternated the situation that investments in science and technology innovation used to be limited to certain sectors and turned to integrate social innovation resources and build innovation clusters, including service chain innovation integrating production, study, research and finance, upgraded industrial chain innovation integrating upper-, middle- and downstream industrial chains, and

[8] Wang Yiming, "China's Green Transformation: Progress and Outlook", *China Economic Report*, Issue 6, 2019.

[9] Wang Yiming, "China's Green Transformation: Progress and Outlook", *China Economic Report*, Issue 6, 2019.

integrated industrial ecological circular chain innovation. Central and local governments simultaneously promote the interactive integration of service chain, industrial chain and industrial ecological chain innovations to effectively boost the commercialization of scientific and technological innovations as well as the iterative upgrading of innovation capacity.

In addition, China is boosting the development of green finance, the construction of a green governance system, and the integration of the green development concept in all aspects and links of the economy, society, culture, science and technology, and management to realize the simultaneous and coordinated transition from "four modernizations" (industrial modernization, urban modernization, agricultural modernization and informatization) to "five modernizations" (industrial modernization, urban modernization, agricultural modernization, informatization and greening) and to stimulate modernization construction to accelerate and transform toward qualitative rather than quantitative development.

Shouguang City, used to be a small county in Shandong Province that grew vegetables for a living, sets up a successful model for green transformation. Starting from the experimental vegetable greenhouses in the 1980s, the "Shouguang model" has been iterating itself. Before 2010, the "Shouguang model" took vegetable greenhouse construction and management technology as core competitiveness and mainly exported products, technology and talent. After 2010, the "Shouguang model" focused on the whole industrial chain from seeds and seedlings to the whole life cycle of vegetable greenhouse construction and management and began to formulate and export vegetable product standards, industrial promotion mechanisms, management and operation systems and other industrial elements. Since 2018, Shouguang has relied on innovations to upgrade the "Shouguang model" and implemented the strategy of "leading throughout the industrial chain" by continuing to consolidate the advantages in greenhouse construction, management, production and operation of the middle-stream chain while focusing on both ends of the industrial chain in the shape of a "smile curve", namely, making comprehensive efforts in the sectors of seed selection and breeding, seedling promotion, functional vegetables, prefabrication, deep processing and brand building to cultivate a USD 13.84 billion competitive and featured industrial cluster with a rational structure and complete chain. In less than 30 years, Shouguang vegetables have become a brand "famous in Weifang, well-known in Shandong, and influential nationwide". According to incomplete statistics, more than half of newly constructed greenhouses in China in recent years have or had "Shouguang elements", meaning Shouguang standards have been applied in 26 provincial prefectures in China. Currently, Shouguang receives more than 2 million visitors every year, and there are more than 8,000 Shouguang technicians across China all year round to guide local vegetable production.[10] In 2000, the first China (Shouguang) International Vegetable Science and Technology Fair was successfully held, and the 23rd Vegetable Fair received more than 30 million exhibitors and visitors from 50 countries and regions, marking "the vegetable capital" of China, Shouguang,

[10] Wang Yulong, "Review on Vegetable Fair: Shouguang Vegetables Lead the Vegetable Industry in China", https://baijiahao.baidu.com/s?id=1730613474560895170&wfr=spider&for=pc.

which is further expanding its influence internationally. Based on integrated green, intelligent and standardized development, Shouguang has gone global.

The 18th China (Shouguang) International Vegetable Sci-Tech Fair, titled " Green, Science, Future"

Source Li Yang, Zheng Pengfei, "The Curtain for the 18 th Shouguang Vegetable Fair Rises, Exhibiting More Than 2,000 Varieties of Vegetables",http://weifang.iqilu.com/wfyaowen/2017/0420/3508632.shtml

2.3 *Liberating and Developing Ecological Productive Forces*

Ecology is a resource, and ecology is productivity. "Lucid waters and lush mountains are invaluable assets."[11] A sound ecology itself is of great economic value and can

[11] Lucid water and lush mountains are invaluable assets", it is an important part of Xi Jinping. Thought on Ecological Civilization. It was first mentioned as a scientific assertion by Xi Jinping, then Secretary of Zhejiang Provincial Party Committee, during an inspection tour in Anji, Huzhou, Zhejiang Province, in August 2005. On October 18, 2017, Xi Jinping, as President of China, pointed out in his report to the 19th CPC National Congress that China should insist on the harmonious coexistence between humans and nature and establish and practice the concept that Lucid water and lush mountains are invaluable assets. On October 12, 2021, at the Leaders' Summit of the 15th Meeting of the Conference of the Parties to the Convention on Biological Diversity, Xi Jinping proposed in his video speech that "Lucid water and lush mountains are invaluable assets. A sound ecology and environment is not just a natural asset but also an economic asset, and it affects the potential and momentum of economic and social development. We need to speed up efforts to

continuously create comprehensive benefits. Nature lays a foundation for human activities, and human civilizations can secure sustainable development only when it is consistent with the development direction of natural ecology. China has to attach more importance to the construction of ecological civilization to truly turn ecology into a productive force and promote the comprehensive development of socialism with Chinese characteristics and the sustainable development of the Chinese nation.

Since the 18th CPC National Congress, governments at various levels in China have upheld the development concept of "lucid waters and lush mountains are invaluable assets" and the principle of balancing ecological, economic and social benefits and have followed the development path of "ecological industrialization", cultivating a large number of eco-industries based on advantages in natural ecological resources. In turn, ecological industries will continuously improve not only regional environmental and ecological quality and stability but also the efficiency of resource utilization, achieving win–win results in economic growth and environmental protection.

In a remote region in western China, there is a province with an eye-catching karst landscape and fragile ecology, Guizhou Province. There used to be inconvenient transportation and a weak economy in Guizhou, as there were sayings that in Guizhou Province, "there would be no three consecutive sunny days, no flat road measuring a kilometer long, and no resident has one Yuan in the pocket". Guizhou used to have the most backwards economy in China, with obsolete production modes and contiguous poor areas. In recent years, under the guidance of the strategic goal of constructing a modernization that ensures the harmony between humans and nature and firmly safeguarding the two bottom lines of development and ecology, Guizhou Province, together with the whole country, has made great efforts in the construction of ecological civilization, environmental protection, and combating desertification while taking local advantage of an excellent ecological environment to vigorously develop a green economy. As a result, Guizhou Province has completely eliminated its absolute property and ushered in a golden decade for development, achieving the leap from poverty to modernization. The beautiful environment and good ecology in Guizhou have attracted an increasing number of Chinese and foreign tourists and talents, giving rise to the development of new businesses, raising the living standard of residents in Guizhou by leaps and bounds, and building a "colorful Guizhou" and "happy Guizhou" that are both prosperous and ecologically beautiful. According to the statistics from Guizhou Province released in 2020, Guizhou Province recorded the third highest economic growth rate during the 13th Five-Year Plan period in China, with an average annual growth rate of 8.5% for gross regional domestic product, 2.8% points higher than the national average. Meanwhile, the green economy developed rapidly in Guizhou, contributing approximately 20% of the gross regional domestic product and leading in the growth rate in China for five years in line. The digital economy in Guizhou contributed the highest growth rate in employment in China for two consecutive years. In addition, the forest coverage rate in Guizhou Province

foster a green way of development and secure a win–win of economic growth and environmental protection, to build a homeland of coordinated advancement of economy and the environment".

reached 61.5%, the proportion of days with good air quality in county towns and cities above the county level reached 99.4%, and the proportion of sections with good water quality of large rivers flowing out of Guizhou was 100%. Guizhou Province recorded improving economic, social and ecological results and was praised as "an example representing the great progress made by the CPC and the country toward the targets since the 18th CPC National Congress". Like Guizhou Province, there have been many successful cases in China as various regions, cities and industries in China achieved high-quality development brought about by ecological productivity, following the road to modernization in which people and nature live in harmony. China, as a responsible major country with greater confidence and vitality, is striving to create a new form of human civilization that ensures harmonious coexistence between humans and nature.

Prosperous and ecologically beautiful "colorful and happy Guizhou"

Source Business News, "Corporate Video of Guizhou Guijiu Broadcasted on CCTV, Demonstrating the Brand of a Maotai-flavour Liquor of Thousands of Years", https://www.163.com/news/article/GKR2EGQ8000189DG.html

3 Upholding and Perfecting the Ecological Civilization System

Modernization requires institutional innovations in advance and systems to escape the process. The construction of an ecological civilization system[12] is an important element in ensuring smooth and sustainable progress toward modernization in which humans and nature coexist in harmony. The Third Plenary Session of the 18th CPC Central Committee proposed to "comprehensively deepen reform, improve and develop the system of socialist with Chinese characteristics, and continue to modernize China's system and capacity for governance," and "centering on building a beautiful China, deepen reform to promote ecological progress and move faster to establish an ecological civilization system, perfect the institutional system for protection of territorial space, the economic use of resources, and eco- environmental protection, to create a new model of modernization that ensures humanity develops in harmony with nature."

3.1 Safeguarding the New Road to Modernization with an Institutional Framework

China is a socialist country, and the system plays a fundamental, global and a permanent role in framing the road to modernization with Chinese characteristics and promoting the sustainable development of humans toward ecological civilization. Only by establishing and improving a mature and sound system can China effectively ensure comprehensive economic and social development. Thanks to the advanced system, China can concentrate the strength for major undertakings!

In 2013, it was proposed in the *Decision of the CCCPC on Some Major Issues Concerning Comprehensively Deepening the Reform* that to build an ecological civilization, it requires a systematic and complete institutional framework for promoting ecological progress, the implementation of the strictest possible source protection system, the system of accountability for ecological and environmental protection and the system of compensation for environmental damage, and the improvement of the environmental governance and ecological restoration system to safeguard the ecological environment with the systems.

In 2015, the Fifth Plenary Session of the 18th CPC Central Committee incorporated "strengthening the construction of ecological civilization and building a beautiful China" in the 13th Five-Year Plan for the first time while deciding to implement the strictest possible environmental protection system, establishing a new

---

[12] The ecological civilization system is a wide concept, referring to a series of rules and systems China establishes to promote the construction of ecological civilization that cover the systems, organizational structure and operating mechanism related to the construction of ecological civilization.

development philosophy including green development, and proposing to improve the institutional framework for promoting ecological progress.

In 2017, the 19th CPC National Congress once again pointed out the need to build a beautiful China by accelerating the reform to promote ecological progress. President Xi Jinping stressed the need to "turn our regulations into rigid and inviolable constraints".[13]

In 2019, the Decisions of the CPC Central Committee on Several Major Issues Concerning How to Uphold and Improve the System of Socialist with Chinese Characteristics and Advance the Modernization of China's System and Capability for Governance put forwards the direction and key tasks to insist on and improve the institutional framework for promoting ecological progress, further clarifying the overall requirements for adhering to and improving the ecological institutional framework and promoting the new journey to modernization (Table 2.1).

3.2 *Setting up Multiple Pillars for the Institutional Framework for Ecological Progress*

The term "four beams and eight pillars" originates from a traditional Chinese building structure, which relies on four beams and eight pillars to support the whole building, so the four beams and eight pillars represent the main supporting structure of a building or a system. According to the latest CPC theoretical achievements, the "four beams and eight pillars" are used as a metaphor to emphasize the need to establish a basic framework for the reform to promote ecological progress.

In 2015, the Guidelines of the State Council and the CPC Central Committee on Accelerating Ecological Civilization[14] and the Integrated Reform Plan for Promoting Ecological Civilization[15] proposed building an institutional framework, which is composed of a system of property rights for natural resource assets, a system for development and protection of territorial space, a spatial planning system, a system of total resource consumption regulation and comprehensive conservation of resources, a system for payment-based resource consumption and compensating conservation and protection efforts, the environmental governance system, the market system for environmental governance and ecological preservation, and the ecological conservation performance evaluation and accountability system, known as the "four beams and eight pillars" of the institutional framework for promoting ecological progress.

---

[13] Xi Jinping, *Xi Jinping: The Governance of China, Volume III*, Foreign Languages Press, 2020, p.363.

[14] "Guidelines of the State Council and the CPC Central Committee on Accelerating Ecological Civilization", *People's Daily*, May 6, 2015, p.1, p.12.

[15] "The State Council and the CPC Central Committee Printed and Issued the Integrated Reform Plan for Promoting Ecological Civilization", *People's Daily*, September 6, 2015, p.14.

Table 2.1 Changes in the efforts to accelerate development of an ecological institutional framework

	2013	2015	2019
Main sectors	1. Improving the system of property rights for natural resource assets and regulatory system for the use 2. Drawing red lines for protecting the ecosystems 3. Implementing the payment-based system for resource use and compensation for ecological conservation 4. Reforming the management system for eco-environmental protection	1. Improving the system of property rights for natural resource assets 2. Developing a system for the development and protection of territorial space 3. Developing a spatial planning system 4. Improving the system of total resource consumption regulation and comprehensive resource conservation 5. Improving the payment-based system for resource use and compensation for ecological conservation 6. Developing and improving the environmental governance system 7. Improving the market system for environmental governance and ecological conservation 8. Improving the ecological conservation performance evaluation and accountability system	1. Implementing the strictest possible systems for environmental protection 2. Developing a comprehensive system for high- efficient resource use 3. Improving ecological protection and restoration system 4. Strictly enforcing the eco-environmental protection accountability system

Data source Decision of the CCCPC on Some Major Issues Concerning Comprehensively Deepening the Reform adopted at the Third Plenary Session of the 18th CPC Central Committee, the *Integrated Reform Plan for Promoting Ecological Civilization* in September 2015, *Decisions of the CPC Central Committee on Several Major Issues Concerning How to Uphold and Improve the System of Socialist with Chinese Characteristics and Advance the Modernization of China's System and Capability for Governance* in October 2019

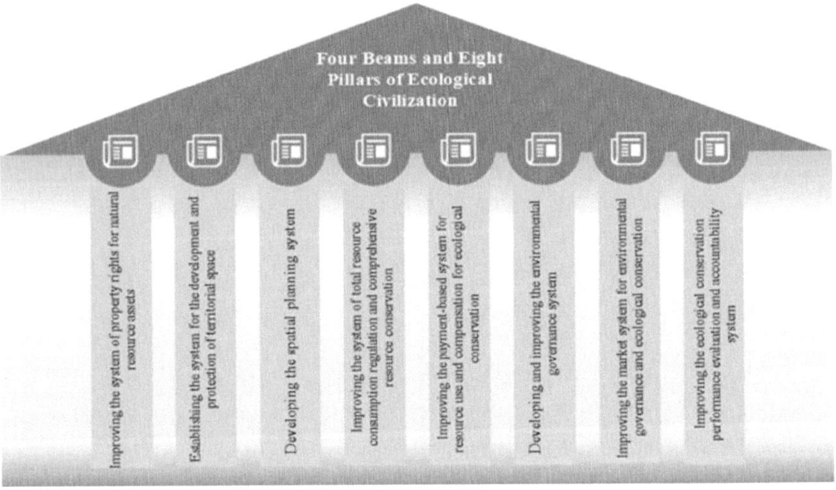

"Four Beams and Eight Pillars" of institutional framework for ecological civilization

The development of these eight systems is designed mainly around the reform in three major sectors: natural resource asset management, natural resource regulation, and ecological environmental protection. Among them, the development of the management systems for natural resource assets mainly includes the system of property rights for natural resource assets, the payment-based system for resource use, the compensation system for ecological conservation, the system of property rights trading, and the system of independent supervision and law enforcement; the development of the systems for regulating natural resources mainly includes the system of spatial planning and use regulation, the red line system of ecological conservation, and the system for auditing outgoing officials' management of natural resource assets; and the development of the systems for ecological environmental protection mainly includes the system of environmental governance and ecological restoration, the system for government purchasing third-party services and conservation concession, the system of environmental offense reporting, the compensation system for environmental damage, the system for regulating total emissions of enterprises and institutions, and the system of lifelong accountability for environmental damage. Five years later, these systems were basically formed through the establishment of four national ecological civilization pilot zones (Fujian National Ecological Civilization Pilot Zone, Jiangxi National Ecological Civilization Pilot Zone, Guizhou National Ecological Civilization Pilot Zone and Hainan National Ecological Civilization Pilot Zone) and will safeguard the construction of ecological civilization in China and further promote modernization that ensures harmony between humans and nature.

3.3 *Promoting Modernization of China's Governance Capacity*

The institutional framework for promoting ecological progress is an important component of China's governance system.[16] The improvement of the governance system is directly related to the modernization of governance capacity, which is a grand project. Therefore, piecemeal adjustments do not work, nor do fragmented repairs. The modernization of governance capacity requires a comprehensive and systematic reform and improvement and the interaction and integration of reforms and improvements in various fields to form the population effect and achieve the overall effect.

First, the reform to promote ecological progress should focus on consolidating the foundation and constructing a framework. In the new era, the comprehensive promotion of reform now emphasizes the improvement, integration, synergy and effectiveness of the systems. The *Decisions of the CPC Central Committee on Several Major Issues Concerning How to Uphold and Improve the System of Socialists with Chinese Characteristics and Advance the Modernization of China's System and Capability for Governance* outlined the institutional framework for promoting ecological progress into four aspects: ecological environmental protection systems, systems for high-efficiency resource use, ecological protection and restoration systems, and ecological environmental protection accountability systems, clearly defining the general requirements for upholding and improving the institutional framework for promoting ecological progress.

The first are the ecological environmental protection systems, which include upholding the harmonious coexistence between humans and nature, insisting on respecting, accommodating and protecting nature, and improving the eco-environmental protection system that covers prevention at source, regulation over operations, compensation for damage, and accountability; accelerating the establishment and improvement of the system for territorial spatial planning and coordinated regulation on the use, coordinating the work on drawing and implementation of spatial control boundaries for the ecosystems (protecting redlines), permanent basic cropland and urban development, as well as various marine area protection lines, and improving the functional zoning system; improving the legal system and policy orientation for green production and consumption, developing green finance, encouraging market-oriented green technical innovation, and more consciously advancing green, circular and low-carbon development; building a mechanism of stationary pollution source regulation centering on pollutant emission permission system, improving the mechanism for cooperation within a region in pollution prevention and control and the integrated mechanism for pollution prevention and control both on land and

[16] China's governance system is an institutional system for managing the country under the leadership of the Communist Party of China and a set of closely interconnected and mutually coordinated national systems composed of institutional arrangements, laws and regulations covering various fields such as economy, politics, culture, society, ecological civilization and Party building.

at sea; strengthening the prevention and control of agricultural and rural environmental pollution; improving the system of eco-environmental protection laws and the judicial system for law enforcement.

The second are the systems for the highly efficient use of resources, including strengthening the rule of law, standards, standardization, and IT application in the unified natural resource ownership determination and registration, improving the property rights system for natural resources, implementing the payment-based system for resource use, and carrying out the system of total consumption regulation and comprehensive resource conservation; improving the policy system for resource conservation and intensive recycling; universally carrying out the system for waste sorting and recycling as useful resources; stimulating the energy revolution to build a clean, low-carbon, safe and efficient energy system; improving the system for marine resource development and protection; accelerating the establishment of a unified system for surveying, assessing and monitoring natural resources; and improving the system for regulating natural resources.

The third are the ecological protection and restoration systems, including carrying out integrated ecological protection and restoration for mountains, rivers, forests, fields, lakes and grasslands, strengthening the protection of natural ecology such as forests, grasslands, rivers, lakes, wetlands and oceans; strengthening the protection of important ecosystems to ensure sustainable use, developing a national park-based nature reserve system, and improving the national park system for environmental protection; intensifying the ecological protection and systematic management of major rivers; carrying out large-scale national greening campaigns, accelerating comprehensive control of soil erosion and desertification, protecting biodiversity, and solidifying national ecological security shields.

The fourth are the eco-environmental protection accountability systems, including developing a system for evaluating and assessing ecological development targets, strengthening the management of binding limits on environmental protection, natural resource regulation, energy conservation and emission reduction, strictly specifying the primary responsibility of enterprises and the supervision responsibility of governments; auditing outgoing officials' management of natural resource assets; conducting comprehensive administrative law enforcement for eco-environmental protection, implementing the system of central inspection on eco-environmental protection; improving the ecological environmental monitoring and evaluation system and the system for ecological and environmental public interest litigation, carrying out the system of compensation for ecological conservation and for environmental damage, and putting into force a lifelong accountability system for ecological and environmental damage.

The achievements made by China have fully proven that modernization with Chinese characteristics is full of vitality and has a promising future. The new road to modernization with Chinese characteristics means a new journey to modernization that ensures harmonious coexistence between humans and nature, and it propels development toward a new form of human civilization, providing a new model for modernization in most developing countries.

Chapter 3
The Development of Ecological Economy

Minjun Shi and Hongbo Chen

In China, it is believed that the development of the ecological economy is a feasible path to coordinate ecological protection and economic development, and it consists of two aspects: first, the integration of industries and ecosystems, which means ecological transformation of traditional production processes, introduction of new environmentally friendly technologies, strengthening resource recycling, improving energy and resource utilization efficiency, and reducing waste emissions; second, ecological industrialization, which means revitalizing ecological resources, turning or integrating ecosystem services into primary, secondary and tertiary industries and realizing their economic value through market-oriented means. For decades, China has made many useful explorations in theory, policy and practice.

1 The Development of Ecological Economy: An Inevitable Choice

1.1 Theoretical Explorations

First, China gained knowledge of the eco-economy from these theories in the West. In the 1970s, faced with the gradually depleting energy resources and increasing environmental pollution brought about by mass industrial production, Western societies began to think about how to balance the relationship between economic and social development and resources and the environment to achieve sustainable development.

M. Shi (✉)
School of Public Affairs, Zhejiang University, Hangzhou, China

H. Chen
School of Applied Economics, University of Chinese Academy of Social Sciences, Beijing, China

K. Yang (ed.), *The Construction of Ecological Civilization in China*, China Insights,
https://doi.org/10.1007/978-981-99-6515-1_3

Therefore, the concepts of "steady state economy", "spaceship economy" and "zero-growth 'global equilibrium'" came out one after another, all of which believed that as a subsystem of the Earth's ecosystem, the economic system must maintain a low or zero growth and low level of consumption to achieve the main purpose of keeping wealth and human stocks constant and sufficient to maintain a good life in the long run to secure that the fluxes that are enough to support these stocks are at a low rather than high level and always within the regenerative and absorptive capacity of the ecosystem. Nevertheless, for developed countries, where there is abundant material wealth and a high material standard of living, "steady state economy" and "zero growth" may be viable options, while for most developing countries in which there is no sufficient food and clothing yet, low or zero growth is hardly acceptable for local residents. Economic growth is an important consideration that governments in developing countries must take into account, meaning that they must find a new way to balance economic growth with resources and the environment. Chinese scholars began to study ecological economics in the 1980s, when they proposed gradually developing Chinese ecological economics, suggesting that of ecological balance and economic balance, ecological balance should be placed at the dominant position since the disrupted ecological balance would finally lead to economic losses. In recent years, ecological economics has been defined as an economic system that can "maintain economic growth that is 'ecologically neutral'". Specifically, ecological neutrality means that human economic activities have a low impact on the ecosystem so that they do not cause damage to the structure and function or affect the long-term stability of the ecosystem. The ecological economy follows the laws of ecology and economics and can maintain a high growth rate to meet the growing needs of the people for a better life without affecting the stability of the ecosystem.[1] To achieve "ecologically neutral" economic growth, the key is to maximize the efficiency and effectiveness of an ecological economy. On the one hand, it needs to integrate industries with ecosystems to improve ecological economic efficiency, namely, achieving maximum economic benefits with minimum resource consumption and environmental pollution to protect "lucid waters and lush mountains" while creating "invaluable assets". On the other hand, it needs to commercialize ecosystem services through ecological industrialization, turning "lucid waters and lush mountains" into "invaluable assets".

[1] Chen Hongbo, "Theories and Practices on Building Ecological Economic System", *Studies on the Socialism with Chinese Characteristics*, No. 4, 2019.

1.2 A Dilemma in Reality

How can the relationship between ecological protection and economic growth be addressed? It is a question easy to answer but difficult to solve. In the early 1980s, when China just carried out the policy of reform and opening-up, there was a backwards economy and underdeveloped industries in China, and the incomes and living standards of the people were far lower than those of the developed countries. In 1980, per capita GDP in China was only USD 431 (in constant US dollar value in 2015, the same below), while per capita GDP in the United States, Japan, Germany, the United Kingdom and France were USD 31,161, USD 19,334, USD 23,766, USD 23,977 and USD 23,572, respectively, for the same year, and 72, 45, 55, 56 and 55 times that of China. The most urgent and important task facing China at that time was to develop the economy and improve the living standards of its people. The story that followed is well known to the world, as China maintained high-speed economic growth for nearly four consecutive decades and created an economic miracle. By 2021, China recorded a per capita GDP of USD 11,188, compared to USD 61,280, USD 35,278, USD 42,527, USD 46,209 and USD 38,210 of the United States, Japan, Germany, the United Kingdom and France, respectively, in the same year, which were 5.5, 3.2, 3.8, 4.1 and 3.4 times that of China, respectively, indicating that the gap between China and those developed countries was narrowed significantly.

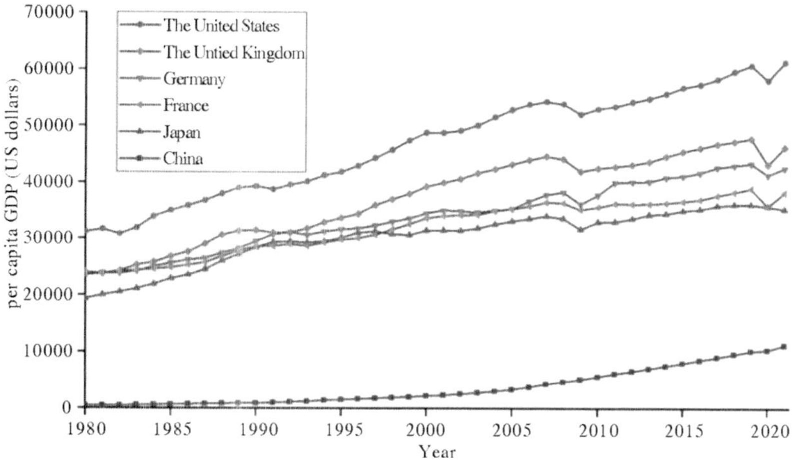

Comparison of percapita GDP between China and major developed countries from 1980 to 2021

Data Source The official website of the World Bank (in constant US dollar value in 2015)

High-speed economic growth in China also brought hikes in energy and resource consumption and pollutant emissions. In 2010, the total energy consumption in China reached 3.6 billion tons of standard coal, surpassing that of the United States and making China the largest energy consumer in the world. In 2019, China consumed energy equal to 4.875 billion tons of standard coal, 1.35, 7.26, 10.1, 16.9 and 13.3 that of the United States, Japan, Germany, the United Kingdom and France, respectively. In 1983, China consumed 118 million tons of iron ore, surpassing Japan (109 million tons), becoming the number one consumer of iron ore globally. In 2017, iron ore consumption in China reached 2.3 billion tons, 53, 18, 59, 237 and 142 times that of the United States, Japan, Germany, the United Kingdom and France, respectively. In 2011, the emissions of sulfur dioxide, nitrogen oxides and particulate matter, as well as the chemical oxygen demand in China, reached 22.18, 24.04, 12.79 and 25 million tons, respectively. The rapidly growing energy and resource consumption and increasingly severe environmental pollution have made economic and social development unsustainable.

In addition, China is still a developing country, with a per capita GDP just exceeding the world average in 2021, ranking 75th among more than 200 countries and regions worldwide. Therefore, maintaining relatively high-speed economic growth is bound to be and must be an important target for China for a long time to come. Economic growth or environmental protection? China faces a dilemma, and it has to foster a new way of development to achieve "win–win" economic results growth and environmental protection.

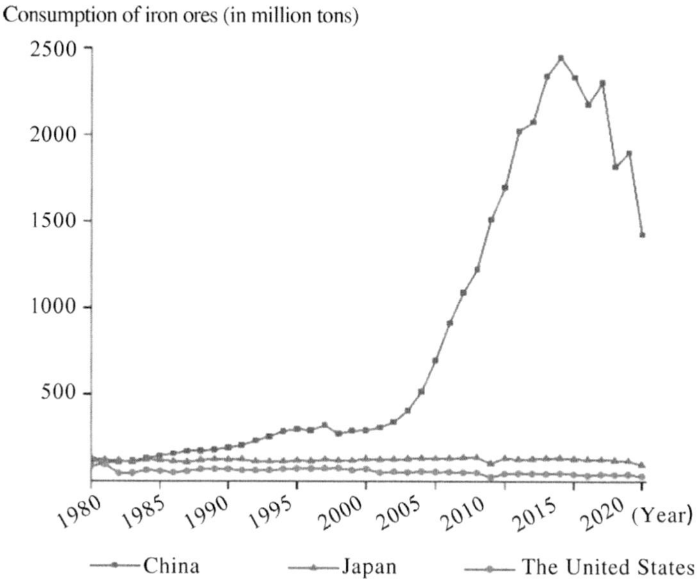

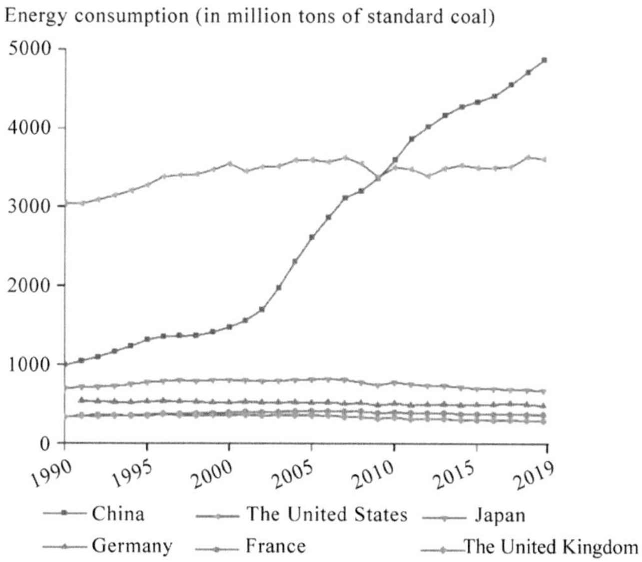

Comparison of iron ore and energy consumption between China and major developed countries

Data Source (1) The data on China's energy consumption were from the *China Statistical Yearbook 2021*, and the data on other countries were from the official website of the U.S. Energy Information Administration; (2) the data on iron ore consumption were from the official website of the World Steel Association

1.3 Evolving Political Consensus

Chinese decision-makers have made active and arduous explorations on coordinating economic development and the environment. In September 1995, the Fifth Plenary Session of the 14th CPC Central Committee formally proposed fundamentally transforming from an extensive to an intensive economic growth mode, indicating that China had recognized that an unsuitable economic growth mode was the root cause for the contradiction between economic growth and the environment. In 1996, China proposed that the essence of environmental protection is to protect the productive forces, showing that China had been aware that the environment is also a productive factor, and it would be possible to resolve the contradiction between economic growth and the environment. In November 1997, China established sustainable development as a national strategy, defining the general policy for dealing with the relationship between economic growth and the environment. In November 2002, the 16th CPC National Congress decided to accelerate the pace of transforming traditional industries and "find a new road to industrialization of high sci-tech content, high economic profitability, low resource consumption, less environmental pollution and fully playing the advantages of human resources, further clarifying the specific method of coordinating economic growth and the environment". It was in this policy paper that China made a political statement that is closest to the definition of ecological economy. In October 2007, the 17th CPC National Congress mentioned ecological civilization for the first time. In November 2012, the 18th CPC National Congress made building an ecological civilization a major component of the "five-sphere integrated" plan for socialism with Chinese characteristics, attaching unprecedented importance to the coordination of economic and social development with the environment. *Guidelines on Accelerating Ecological Civilization* issued in 2015 made it clear that green, circular and low-carbon development is the basic way to achieve ecological progress, further mapping out the method of coordinating economic and social development with the environment. In November 2015, the Fifth Plenary Session of the 18th CPC Central Committee put forwards the "new development philosophy emphasizing innovative, coordinated, green and open development for all". The 19th CPC National Congress in November 2017 proposed that "man and nature form a community of life", developed and practiced the concept of "lucid waters and lush mountains are invaluable assets", and pointed out that ecological productivity is the most fundamental and creative productive force, and a good eco-environment can generate more economic benefits. At the National Conference on Environmental Protection in May 2018, President Xi Jinping formally proposed "to establish an ecological economic system aimed at integrating industries and ecosystems". By then, the theoretical system for China's ecological economy had taken shape, setting the direction for policy and practice.

2 Making Production Cleaner

To improve the efficiency of energy and resource utilization and reduce pollutant emissions in the production process, China encourages enterprises to upgrade their production process by implementing clean production. As stated in the *Law of the People's Republic of China on the Promotion of Clean Production*, which was enacted on June 29th, 2002, enterprises should, through "continuously improving designs, the use of clean energy and raw materials, the use of advanced technology and equipment, improvement of management and comprehensive utilization of such measures, reduce pollution from the source, improve resource utilization efficiency, reduce or avoid generation and emission of pollutants in the process of production, service and product use in order to reduce or eliminate the harm to human health and the environment". Later, the relevant state departments formulated and introduced a series of supporting policies and systems, including *Opinions on Accelerating the Promotion of Clean Production, Interim Measures for Clean Production Audit, Regulations on Clean Production Audit Procedures in Key Enterprises, Notice on Further Promoting Clean Production in Key Enterprises*, and *Interim Measures for Special Funds of the Central Finance for Clean Production.* On July 1st, 2012, the amended *Law on the Promotion of Clean Production* was put into force, indicating that the strategy of prevention at the source and regulation over the whole process was integrated into the comprehensive strategy of economic development.

Based on the promotion of clean production, China has proposed environmentally friendly production methods, promoted the circular economy model following the principle of "reducing wastes, and recycling wastes as resources for reuse", green product identification, and so on, and gradually advanced the construction of ecological industrial parks, green recycling parks and green industrial parks throughout the country following the procedure of "clean production, circular economy, sustainable development, and green development", significantly improving the environmental performance in various industries and sectors. For example, while the production value continued to grow, the emissions of industrial solid wastes were effectively reduced to 560,000 tons in 2015, from 31.86 million tons in 2000. Phenomena such as indiscriminate dumping of industrial wastes were generally curbed, and the comprehensive utilization rate of solid wastes increased from 45.9% in 2000 to 60.2% in 2015.

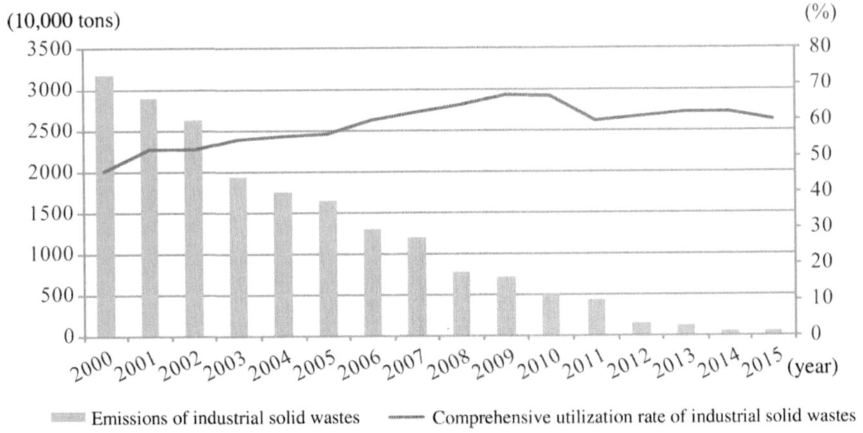

Emissions and comprehensive utilization rate of industrial solid wastes in China from 2000 to 2015

2.1 Up-and-Downs of a Cement Factory

Over the past sixty years since it was founded in 1956, Jidong Cement Tongchuan Co., Ltd. has experienced a technical change from wet to dry production and expansion, restructuring and listing, bankruptcy reorganization and major assets reorganization, but now it has constructed an aggregate project with an annual production capacity of 6 million tons and ranks third in China and fifth in the world in terms of cement production capacity. This cement producer used to be a large energy consumer and pollutant emitter in Tongchuan, causing a severe impact on local environmental quality. In 2021, it invested USD 276.8 million and built a new 10,000-ton cement production line while shutting down four cement clinker production lines and one aggregate production line. In addition, it set up the first cement intelligent laboratory in China, pioneered the use of AGV carts to transport samples in the cement industry, and innovated in or invented a new clinker sampling and sample preparation system, nitrogen pulverized coal transportation system, and smart bagging and shipping system, realizing capacity upgrades.

Digitalization and green transformation also contributed to the enterprise's transformation. Through digitization, the enterprise raised the digitization rate of equipment in the new production line to over 95%, improved the stability of the kiln and mill expert control system by more than 15% after it was put into operation, and lifted the accuracy of the digital prediction model for each process section, equipment and critical quality to above 60%, realizing full automation of material management, quality control, production management, equipment operation and maintenance of the production line. Meanwhile, thanks to artificial intelligence (AI), which enables self-learning and self-improvement via big data, the enterprise improved its overall corporate management. In terms of green development, the new production line adopted a six-stage cyclone preheater and online spouted calciner, with an annual production capacity of 3.1 million tons of clinker, and it has a complementary 15 MW low-temperature waste-heat power generation system. In a digital twin factory that gathers data from various sectors, in real time, the enterprise can analyse production, security, energy-saving and emission-reducing data and manage to reduce energy consumption by 16.3% on average compared to before the upgrade, namely, overall power consumption per ton of cement down to 68.05 kWh and energy consumption to 92.46 kg standard coal. In addition, the comprehensive energy consumption was improved by 20%. The enterprise also realized ultralow controlled pollutant emissions, 100% recycling of solid wastes and zero discharge of wastewater. The new production line cuts the emissions of key pollutants significantly, for example, reducing the emissions of particulate matter to 24.31 tons, nitrogen oxide to 79.64 tons and sulfur dioxide to 8.76 tons, 85%, 93% and 63% lower than before the renovation, respectively. Meanwhile, compared with the conventional standard line, this production line has a doubled single line capacity, and the productivity per worker was increased by three times, both reaching an international advanced level.

Overview of Jidong Cement Tongchuan Co., Ltd.

Source Wang Feng, "With an Investment of Nearly USD 692 Million, Two Cement Clinker Production Lines with a Daily Capacity of 10,000 Tonnes Each Launched Operation on the Same Day in Tongchuan!", https://www.sohu.com/a/474688073_121124549

Control center of Jidong Cement Tongchuan Co., Ltd.

Source Cheng Honglan, "Tongchuan: Industrial Structure is being Optimized, Transformation and Upgrading Accelerates", http://szb.tcrbs.com/tcrb/20220219/html/content_20220219003001.htm

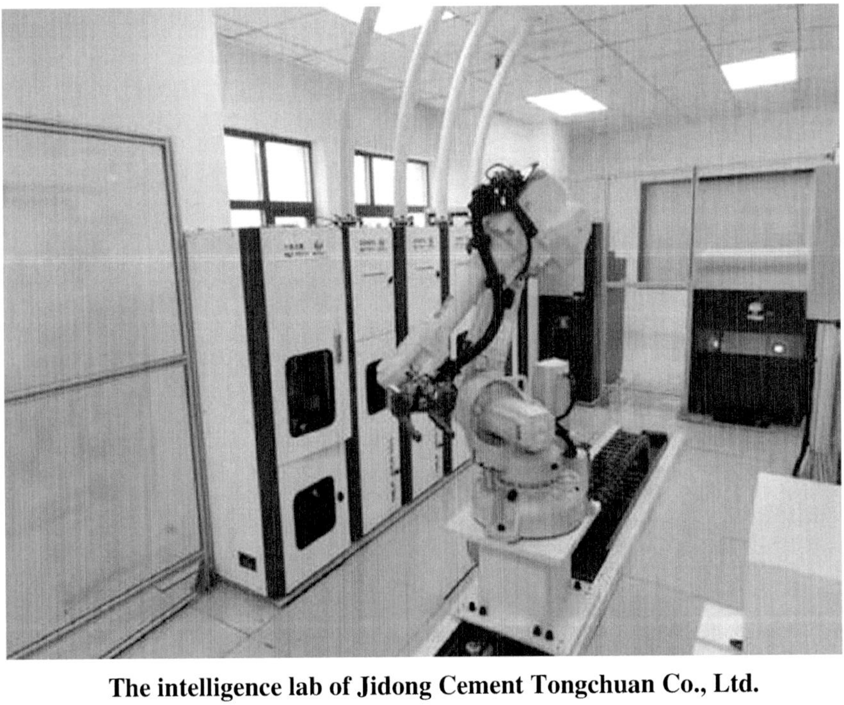

The intelligence lab of Jidong Cement Tongchuan Co., Ltd.

The case indicated that automatic, digital, intellectual and green upgrades can help enterprises achieve zero accidents and low-cost operations, reduce labor intensity, improve the happiness index of workers, lower energy consumption and carbon emissions, optimize comprehensive management and economic indexes, and gradually form a new business mode integrating technology, products and business and supporting industrial derivation.

2.2 For the "Blue Sky" Over Chemical Industrial Parks

The Hangzhou Bay Shangyu Economic and Technological Development Zone (HSEDZ) is located in Shangyu District, Shaoxing City, Zhejiang Province, on the south coast of Hangzhou Bay, and forms an economic growth pole for Shangyu District. In 2016, HSEDZ realized a total industrial output value of more than USD 13.84 billion, accounting for approximately 50% of the total industrial output value in Shangyu District. The fine chemical industry is the pillar industry of the development zone, and the chemical industry contributed 60% of the total output value of the development zone in 2016, and enterprises of dye (paint) and intermediate products in particular accounted for 30% of the total output value in the HSEDZ.

Before 2004, chemical enterprises in the development zone emitted waste gas directly, resulting in severe air pollution that seriously impacted neighboring residents, and the problem of not-in-my-yard could not be solved for a long time. Hangzhou launched special regulation plans for waste gas in 2004 and has carried out two rounds of special remediation actions for the chemical industry since 2007, covering 811 chemical enterprises. In 2012, Hangzhou regulated the lead-acid battery industry and four industries of chemical printing and dyeing, papermaking and so on in 2013 and 2014, while it initiated special projects to scientifically control waste gas between 2015 and 2017.

To promote the fundamental transformation of chemical enterprises, the development zone launched the "Blue Sky" campaign in 2017 to eliminate backwards production capacity and "low-tech, small and scattered" chemical enterprises. First, it promoted the shift from end treatment to source reduction and encouraged enterprises to adopt clean production processes and increase investment in environmental treatment facilities. In the development zone, Zhejiang NHU Company implemented the clean production of "reducing emissions at the source, regulation over operation and treatment at the end" and shifted the focus of environmental governance from the end of pipe to the stages of production and R&D design; Menovo Pharmaceutical invested USD 3.46 million in 2017 to rebuild the sewage station, boosting the designed sewage treatment capacity of the new sewage station to 1,000 tons per day, while constructing a pretreatment system in the production workshop to strengthen the control of pollutant emissions at the source with falling film, acid and alkali absorption and other measures. Second, from shutting down outdated production capacities, the HSEDZ turned to focus on gradually strengthening environmental governance and regulation. In 2017, 119 chemical enterprises in the development zone invested a total of USD 1.87 million, built an intelligent rainwater collection system, and equipped intelligent monitoring facilities outside the drain holes of clean water, which can close the valves on days without rain, quickly open the valves when it rains heavily, and automatically and synchronously start the sampling devices to monitor the water quality discharged from the drain holes of clean water. Through strict rainwater and sewage diversion control and real-time monitoring and supervision, the development zone effectively reduced the concentration of COD in major rivers. Third, the HSEDZ coordinated efforts to strengthen environmental regulation and to guide industrial transformation, boosting orderly industrial structural adjustment and optimization. In accordance with the principle of "clustering and upgrading some, merging and reorganizing some, and shutting down and eliminating some enterprises", the development zone took measures to guide the transformation and upgrade of the chemical industry. By 2017, 34 chemical enterprises in the HSEDZ had been ordered to suspend production for rectification, 26 low-efficiency enterprises had switched to nonchemical businesses, and 13 enterprises had been merged and reorganized, while 80 or more enterprises had carried out another round of comprehensive rectification actions to improve production safety, environmental protection, fire protection and so on.

Strict environmental governance effectively promoted clean production in chemical enterprises. With huge investments in environmental protection facilities for source reduction and end-of-pipe treatment, enterprises upgraded production processes and safety management, saving production and environmental costs. For example, Zhejiang NHU Company created economic benefits of USD 156,120 a year with the project of sludge reduction, containment and continuous treatment, of USD 463,640 from the incinerator upgrading project, and of USD 420,740 from the programmed operation of intermittent processes. Zhejiang Longsheng Group (Lonsen) used to neutralize the waste high-concentration mother solution with lime for dye production, producing a large amount of sludge (hazardous wastes). The new MVR facility injects ammonia water into highly acidic wastewater, which can reduce the sludge by 82.5% and the sludge treatment costs with the land filling method while obtaining additional income from the by product of ammonium sulfate. If the original sewage treatment method is still in place, it would require an expenditure of nearly USD 13.84 million for Lonsen to bury sludge. In other words, the new facility helps Lonsen cut the overall costs of sludge treatment by USD 11.36 million or so. In addition, strict environmental regulations have helped local chemical enterprises improve their competitiveness in the market and profitability. In the first ten months of 2017, HSEDZ recorded a double-digit increase in the total output value compared to the same period a year ago.

2.3 *Licence System in the Textile City*

The textile printing and dyeing industry is the pillar industry in Keqiao District, Shaoxing City, Zhejiang Province. Over more than 40 years, the textile industry in

Keqiao District formed a complete textile industry chain covering all aspects from PTA and chemical fiber to weaving, printing and dyeing, then to design, clothing and home textiles. Keqiao District gathers approximately 1/3 of the total printing and dyeing production capacity in China, and nearly 1/4 of global textile products are traded in Keqiao, making this "Textile City" a good example demonstrating the great achievements of the reform and opening-up in Zhejiang. As the textile printing and dyeing industry seriously pollutes the environment, since 2013, Keqiao District has implemented strict environmental governance measures, especially the innovative "licensing" model.

First, this "Textile City" developed a comprehensive evaluation system based on contributions per *mu* (a unit of area, = 0.0667 hectares or 0.1647 acres) of land to attract textile printing and dyeing enterprises to settle down in local industrial parks and enjoy good infrastructure and a supporting environment. Meanwhile, Keqiao District adopted stricter environmental standards to force printing and dyeing enterprises to eliminate backwards capacity, save energy, reduce consumption, conserve water, decrease emissions, and innovate in science and technology. Furthermore, it took back the land allocated to enterprises that could not meet the requirements of environmental governance for reallocation, thus not only realizing industrial clustering and intensive utilization of land but also promoting industrial structural adjustment and environmental protection.

China textile city in Keqiao District, Shaoxing City, Zhejiang Province

Source Zhejiang Convergence Media, "Topping One Billion US Dollars! The Export Trade Value from Zhejiang Keqiao Textile City through Market Purchasing Made New Records", http://k.sina.com.cn/article_7505202169_1bf584bf902000xgfg.html

Second, Keqiao District introduced guiding environmental standards and a negative list system for the treatment of three wastes. While attracting printing and dyeing enterprises, the government in Keqiao District made positive and negative lists on production equipment, process technology and treatment of three wastes and applied strict and high standards for new plants, newly purchased equipment and process technology routes, encouraging textile printing and dyeing enterprises to widely adopt new green and environmentally friendly equipment and advanced technology. For example, Keqiao District enacted the first group standard for the textile painting and dyeing industry in China, *Green Printing and Dyeing General Requirement* (T/SKBJ 001-2017), which clearly defined the basic standard for comprehensive energy consumption per hundred meters of chemical fabric as 28 kg standard coal or less, and industry-leading standard of 26 kg standard coal or less. Keqiao District also organized enterprises to actively participate in the revision of the water pollutant emission standards for the textile industry (GB4287-2012) and became a standard-setter involved in formulating standards for sewage, exhaust gas and other pollutant emissions, as well as green printing and dyeing specifications for the textile industry.

Third, Keqiao District innovatively established the "licence" system, meaning that more than 2,000 sets of setting machines in all 109 printing and dyeing enterprises are "licenced", equipped exhaust gas collection devices and real-time monitoring systems, and fully realized the diversion of rainwater, sewage and wastewater based on water quality and wastewater discharge to the public sewage pipe network system. After all, Keqiao District achieved a high collection rate for exhaust gas (more than 90%), oil smoke (more than 60%), and total particulate matter (more than 80%) and improved regional air quality. In terms of economic benefits, the added value per meter of printing and dyeing fabric was increased by more than 15%, and the average tax revenue per *mu* was lifted from USD 18,643 to USD 39,818.

3 Making the Ecological Environment More Valuable

3.1 Exploring the Path to Realize Ecological Value

Ecosystems are the material basis for human survival, providing humans with various productive and living factors, such as food and tourism resources, while playing a vital role in regulating climate and environmental quality, maintaining soil fertility and nutrient cycles, and sustaining biodiversity. It is the various nonstopping services from the ecosystems that secure humans with constant supplies of clean water, fresh air, fertile soil and pleasant climate and that support sustainable economic and social development. Thus, ecosystem services contain a huge value of ecological services. According to Costanza et al., it was estimated that the value of global ecosystem services in 2011 reached USD 125 trillion, 82% higher than the global gross national product of USD 68.85 trillion in the same year.[2] However, the huge potential value of ecosystem services does not always turn into real economic value. Although there is a good eco-environment in many places, local residents are struggling for a living, as they have low economic incomes and lead a poor life. Therefore, how to develop ecosystem services into ecological industries and transform ecological values into economic benefits, meaning turning "lucid water and lush mountains" into real "invaluable assets", is a challenge, and so is the road of eco-economy that the Chinese government and enterprises are seeking.

The integration between industries and ecosystems started with agriculture. In 2002, the Ministry of Agriculture selected ten representative eco-agricultural models and supporting technologies for promotion, including the "four-in-one" model (a household-based eco-agricultural model integrating biogas production, waste treatment, planting and feeding) and its supporting technologies, which gradually developed into comprehensive models integrating several new technologies and achievements and then a systematic eco-agriculture combining various models and technologies. Subsequently, eco-industry and eco-tourism emerged and boomed, forming an eco-industrial chain with organic linkages between primary, secondary and tertiary industries. At present, the integration between industries and ecosystems is mainly to tap the market value of ecological resources, transform ecological services and improve the quantity and quality of the supply, and increase the added value of ecological products and services through the organic integration of advanced and scarce production factors, including the combination of modern agricultural technology and local characteristic breeding industry, advanced management and operation experiences and ecological tourism, e-commerce and household agricultural products sales, etc. In particular, in remote areas with a small population and rich ecological resources, the added value of ecological resources can be rapidly realized or increased by introducing capital, technology and management services, improving the scale and quality of operation. In some regions, local governments focused on constructing a network of interconnected and interdependent industries, such as developing ecological tourism-related industries to form an industrial complex combining catering, accommodation, production and sales, integrating urban and rural telecommunications, transportation and agricultural and byproduct processing industries to gradually reduce the actual costs of ecological industries, and promoting the extension of the industrial chain and the integration of urban and rural areas to coordinate the targets of realizing ecological values and increasing incomes of urban and rural residents.

In April 2021, the General Office of the CPC Central Committee and the General Office of the State Council jointly issued the *Opinions on Establishing a Sound Mechanism for Realizing the Value of Ecological Products*, calling for diversifying

[2] Costanza R, Groot RD, Sutton P, Ploeg SVD, Anderson SJ,Kubiszewski I, Farber S, Turner RK, "Changes in the Global Value of Ecosystem Services", *Global Environmental Change*, 26, 2014, pp. 152–158.

the modes for realizing the value of ecological products. In addition, it also encouraged governments and enterprises in different regions to explore various models and ways to promote the realization of ecological product values in a scientific and reasonable manner under the premise of strict eco-environmental protection; improve the value of ecological products by relying on the unique regional natural endowment; and adopt original ecological cropping-breeding modes such as artificial breeding and natural growth, self-reproduction and growth, etc.; scientifically apply advanced technologies for deep processing, expand and extend the industry and value chain for ecological products; take advantage of local high-quality water, air, pleasant climate and other environmental background conditions; moderately develop the digital economy, clean pharmaceutical business, electronic components manufacturing and other environmentally sensitive industries; promote the transformation of ecological advantages into industrial advantages; make use of beautiful natural scenery and historical and cultural relics; and form professional design and operation teams to create ecological tourism development models that integrate tourism, recreation, leisure and so on, on the condition of minimizing artificial disturbances; accelerate the cultivation of market entities involved in ecological product marketing and development; incite the revitalization of stock resources such as abandoned mines, industrial sites and ancient villages; promote the centralized transfer and operation of related resource rights; and coordinate systematic eco-environmental governance and the construction of supporting facilities to enhance the value of educational and cultural tourism development.

3.2 Dramatic Changes in Xiufanggou

The Xiufanggou Coal Mine is located in the upper reaches of the Xiufang River in Tongchuan City, Shaanxi Province. The coalmine started production in 1997, with an annual mining capacity of 900,000 tons. After nearly two decades of mining, the ecology of the Xiufang River Basin was seriously damaged: the lucid water and lush mountains became black and covered by a thick layer of coal power. The water flow in the river weakened and even vanished from time to time, and the water quality deteriorated, causing water plants and animals to die out and birds to fly away. As trucks swept along the road on the banks, black clouds of dust and coal powder were stirred up, covering everything including the yards, roofs, trees, croplands, and riverbed. In other words, the living conditions were harsh. Around 2014, due to sharp drops in coal prices, a great number of coal mines struggled for survival, and many had to suspend production. At that time, the Ministry of Finance, the Ministry of Land and Resources and the Ministry of Environmental Protection issued policies and launched pilot projects for the ecological protection and restoration of mountains, rivers, forests, fields and lakes and for the ecological restoration and transformational development in old mining areas and other regions suffering serious ecological damage with central financial support. The owners of the Xiufanggou Coal Mine, after consulting with the local government, decided to close the coal mine and apply for central financial support to restore the ecology of the Xiufang River and develop tourism.

In 2015, the central and local governments invested USD 38.75 million and carried out three major projects of comprehensive watercourse ecological treatment, protection and comprehensive utilization of water resources, comprehensive control of soil erosion and ecological restoration. The coal mining enterprise invested USD 122.35 million to develop the Xiufanggou scenic area: based on the well-known work of *Travelers Among Mountains and Streams* (《溪山行旅图》) by Fan Kuan, a famous painter in the Song Dynasty, and guided by the artistic realm of Chinese paintings and the landscape described by Fan Kuan, the development project targeted building the first scenic spot in China, reminding visitors of the Song Dynasty, and constructed more than 40 leisure and resort facilities, including the Xishan Painting Museum, XishanYiju Hotel, Xishan Gallery for Sketching from Nature, Xishan Health Preservation Park, Whispering Birds' Forest Wetland Ecological Park, Xishan Street, Changkong Cableway and Xingkong Hotel. When developing this old mine, Xiufanggou Coal Mine followed the principle of "no big demolition nor construction, restoring the old as the old", renovated and upgraded existing buildings and developed industrial tourism: through real objects and tools and coal culture exhibition hall equipped modern light and sound technologies, let visitors intuitively experience the coal mining process underground. In 2017, Xishan scenic spot was designated as a national 4A-level scenic spot, also enlisted among national preferred selected tourism projects, and became a sketching base for the Shaanxi Artists Association in Xishan, Zhaojin Town, an off-campus educational practice base for college students of Xi'an Academy of Fine Arts, and a creative research base for Xi'an Chinese Painting Research Association. Currently, the Xishan scenic spot is a well-known "summer camping resort" and a "famous destination on the Internet" and now receives 300,000 visitors on average a year, contributing a revenue of USD 3.86 million and providing more than 200 jobs. In addition, the annual per capita income of local farmers reached USD 4,982.

Xiufanggou scenic area

Source Culture Art News, "The Number of National 4A-level Scenic Spots in Tongchuan Rose to 22", https://www.sohu.com/a/537622493_556618

3.3 Developing Ecological Products into Brands

Lishui City, located in the mountainous region of southern Zhejiang Province, used to have poor transportation facilities and lagged behind in industrialization and urbanization. However, due to less pollution, there are better and more diverse environmental background conditions, which is conducive to the development of pure natural and original ecological agriculture. However, there were constraints, too, as it was hard to integrate these household-based small and scattered businesses to establish characteristic agricultural product brands, and it lacked an efficient channel to match the supply of high-quality agricultural products with the market demand. In September 2014, under the guidance and promotion of the local government, the Lishui Ecological Agriculture Association established the regional public brand of Lishui Shangeng to promote the development of ecological agriculture.

By fully taking advantage of the environmental background conditions for ecological agriculture, boosting the standardization of ecological agricultural production through an information traceability system, enhancing the value of ecological agricultural products with brand certification, and promoting efficient circulation and brand marketing with the "One Ecology" system, Lishui City continuously makes the brand of Lishui Shangeng larger and stronger. First, Lishui City introduced a standard certification system and developed the first regional public agricultural brand with brand certification in China. In 2018, the brand standard system of "Lishui Shangeng" was implemented, which was composed of standard A (general standard) + standard B (product standard). The first batch of 21 enterprises was certificated by the International Certification Alliance for Lishui Shangeng, and 23 brand certificates of Lishui Shangeng were issued, indicating that Lishui City began to boost standardized brand management with the model of third-party certification. After 2018, Lishui restricted and prohibited the use of three batches of 106 pesticides and fertilizers, taking EU standards as a benchmark, and fully applied a real-name registration system for pesticide purchases to protect the "ecological attributes" of the products. Second, Lishui developed a traceability and supervision system throughout the product life cycle for comprehensive quality assurance. The brand operator of Lishui Shangeng insists on the strategy of "direct supply from the production base, testing before sales and life-cycle traceability", conducts strict testing on the quality of agricultural products, and sets up a four-stage (city, county, township, and enterprises) quality and safety traceability and supervision system for products of nine categories (vegetables, fruits, edible mushrooms and so on). As of June 2018, 1,419 enterprises in Lishui City were covered by two province- and city-based traceability systems, while the city completed 15,465 product testing samples of 851,748 items and compiled 16 operation manuals for chain storage and transportation of agricultural products. The number of anti-counterfeit two-dimensional code labels to be approved and in use reached one million. Third, Lishui City continuously broadens and coordinates online and offline marketing channels to form a marketing system integrating e-commercial and WeChat businesses and physical stores. Based on the "Internet of Things + Big Data", Lishui City built "One Ecology" information service system to connect with the GS1 system, which has a global unified logo, and provide big data services. In addition, Lishui City organized and actively took part in Lishui Ecological Excellence Agricultural Expo, Zhejiang Agricultural Fair, Shanghai (Zhejiang) Brand-name Expo and other brand promotion activities, while making efforts to turn ecological products of the Lishui Shangeng brand into tourism commodities and promoted the construction of sales networks in tourist spots. As of June 2018, 254 agricultural products of the Lishui Shangeng brand were sold through e-commerce platforms, and 173 offline marketing outlets were set up.

By April 2021, the Lishui Ecological Agriculture Association had 629 members, and 390 enterprises in Zhejiang were certified with the Lishui Shangeng logo, with 531 certificates being issued. A total of 1,419 enterprises in Lishui City were covered by two provinces and city-based traceability systems, while the total number of anti-counterfeit two-dimensional code labels to be approved and in use reached 3.20 million. An industrial structure dominated by agriculture has been formed in Lishui, in particular by nine categories of products of edible fungi, tea, fruits, vegetables, medicine, animal husbandry, camellia oleifera, bamboo shoots and fishery. In 2020, the annual sales of Lishui Shangeng agricultural products reached USD 1.5 billion, with an average premium rate of 30%. In 2017, the value assessment of regional public brands for agricultural products indicated that Lishui Shangeng had a brand value of USD 368 million, and it was the most influential regional agricultural brand in China for three consecutive years in 2018, 2019 and 2020 in terms of regional agricultural image and brand category.

丽水山耕
LISHUI SHANGENG
Follow the Laws of Nature,
Enjoy the Purity of Ecology

Logo of Lishui Shangeng

Source "Interpretation of the Regional Public Brand of Lishui Shangeng", https://www.sohu.com/a/332451816_120146688

3.4 A City with Beautiful Water Landscapes is Additionally, Attractive to Investors

Nanping is the birthplace of the Minjiang River, the mother river of Fujian Province, and Wuyi Mountain, a world natural and cultural heritage site, is also in Nanping. Nanping city possesses rich ecological resources in Fujian, as there are lucid water and lush mountains and a forest coverage rate as high as 74.75%. However, due to its topographic features and location, Nanping was once the least developed city in Fujian and ranked last in Fujian Province in terms of total and per capita GDP, meaning it did not transform the advantages of ecological resources into economic benefits.

In 2015, based on the local outstanding eco-environment, Nanping City began to build a city with a beautiful water landscape by comprehensively studying the evolution of the water system in the city and focusing on the watershed area and coordinated relationship among mountains, water and the city. By leaving space for water to flow in and out, Nanping aimed at the natural circulation of water bodies in urban areas. Along watercourses, Nanping City constructed urban facilities to integrate rivers, riverbanks, landscapes, roads, cultural heritages and the city for harmonious coexistence between water and the city and between water and nature while coordinating urban water conservancy and landscape construction to make full use of water to landscape the city and blend the city with surroundings. Meanwhile, Nanping updated the old part of the city with the sponge city concept and targeted "transforming rainwater and sewage mixture for diversion, alleviating the problem of waterlogging and improving environmental quality" and renovated old neighborhoods through the construction of sponge facilities, including rainwater disconnection, pipe network transformation and biotretention measures, to solve problems such as rainwater and sewage mixture and waterlogging, improve the living environment and build ecological and livable homes. Meanwhile, the city carried out biodiversity protection, soil erosion control, water environment protection of river basins, environmental restoration in mining areas and other ecological conservation and restoration projects to regulate livestock, poultry and domestic pollution and illegal sand mining and promote black and odorous water treatment, including comprehensive improvement of thirteen small river basins. In recent years, Nanping City invested approximately USD 511.53 million in river governance and flood control and constructed flood banks of 285 km and revetments of 4.5 km, being praised by the Ministry of Water Resources and the Ministry of Housing and Urban–Rural Development as a city with beautiful water landscapes.

Nanping, a city with beautiful water landscapes

While focused on the construction of water landscapes, Nanping City comprehensively accelerated the development of the new tourism industry, making deep explorations into six new elements of tourism: "business, health-keeping, training, leisure, sports events and adventure" and enriching tourism products such as city sightseeing tours, sports and fitness tours, science and technology tours, health-preserving tours, food tours and rural experience tours. For example, the Water City project in Wuyi Mountain launched the "night tour of Chongyang Creek", and the Water City project in Shunchang built cruises for tourists to Fujin Lake, providing the tourism industry with new ecotourism products. The improvements in environments and public infrastructure have strongly enhanced the attractiveness to investors, and in 2021, Nanping City signed agreements for 164 industrial projects, each with an investment of more than USD 13.84 million. These investment projects raised local land prices. Taking Shunchang County as an example, the county expanded its urban area by 3.6 km^2 due to urban development, realizing an increase in revenue on the transfer of land use rights of more than USD 415.2 million.

Chapter 4
Safeguarding Ecological Security

Wenying Peng and Jingjing Shan

China is a country with a large population, limited land resources, uneven spatial and temporal distributions of water resources, frequent natural disasters and relatively fragile ecosystems. Safeguarding ecological security is not only the most basic and essential guarantee for human survival and economic and social development but also the basic goal and an important part of ecological civilization. Since the founding of the People's Republic of China, the social understanding of ecological security in China has been deepened continuously, evolving from initially keeping the environment clean to strengthening environmental protection, then safeguarding ecological security and promoting ecological civilization, and from focusing on safeguarding national ecological security to calling for "making due contributions to global ecological security" and "building a Shared Future for All Life on Earth". China is gradually establishing the concept of ecological civilization, solidly promoting ecological progress and continuously strengthening ecological security.

1 Continuously Strengthening Ecological Security

The eco-environmental condition of a country or region is profoundly influenced by its natural conditions and human behavior. China has a huge population of 1.4 billion, with a per capita arable land area of only 1.36 mu because there are more mountains and fewer plains in China, far below the world average. China's water

W. Peng (✉)
Capital University of Economics and Business, Beijing, China

J. Shan
Research Institute for Eco-civilization, Chinese Academy of Social Sciences (CASS), Beijing, China

K. Yang (ed.), *The Construction of Ecological Civilization in China*, China Insights,
https://doi.org/10.1007/978-981-99-6515-1_4

resources per capita are 2,000 cubic meters/person-year, only a quarter of the world average, meaning that China is one of the poorest countries in the world in terms of per capita water resources. Under the influence of the monsoon climate, the spatial and temporal distribution of water resources in China is extremely uneven: "more in the south and less in the north, more in the east and less in the west, more in summer and less in winter". Since land is the great mother of all, and water is the foundation of ecosystems, constrained by natural conditions, China has a relatively fragile eco-environmental background, and coupled with the high-intensity development and construction brought by the rapidly developing economy and society since the reform and opening-up, as much as 55% of the territory has a moderately or above fragile ecology. It is a major long-term challenge for the Chinese government and its people to meet the people's ever-growing needs for a better life while effectively safeguarding ecological security in a country with a relatively fragile ecological environment and tight per capita resources.

1.1 The Definition of Ecological Security

In essence, nature is a complex ecosystem, an organically unified whole formed between organisms and between organisms and the environment in a certain time and space. Nature not only provides humans with rich sources of living materials, such as grains, vegetables, fruits and fishery but also colorful sources of production materials, such as oil, timber and minerals. The survival and development of humans depends on nature, and human production and living activities also affect nature. "Ecological security" means that the natural ecosystem on which the earth, a country or a region depends for its survival and development is in a state free from or less threatened by damage, and generally, it includes two aspects. The first is the security of the ecosystem itself, that is, whether the ecosystem itself is damaged, whether its structure is intact, whether its function is sound, and whether its state is stable. The second is the security of human society, namely, whether the products and services provided by the ecosystem can guarantee the needs of human society for safe survival and healthy and sustainable development.

Ecological security has systemic, holistic, dynamic and fundamental characteristics. First, it is systemic. The ecosystem is composed of various elements of nature, including organisms and the environment, while the organisms and the environment, as well various elements, interact with, influence and constrain each other and are closely linked, forming an organic whole with certain functions. Problems in any element or any link of the ecosystem may cause the chain reaction of other elements or links, leading to systemic risks in ecological security. Second, it is holistic. The ecosystem is an inseparable whole, and damage to the regional ecological environment or a subsystem may result in a global eco-environmental issue, thus bringing about overall risks to ecological security. Third, it is dynamic. The ecosystem is influenced by complex factors, and it responds differently according to changing influencing factors. Ecological security is a relatively stable and dynamic balance, and there is no absolute but relative security. Fourth, it is fundamental. The national security system includes political, homeland, economic and social security, etc., and ecological security is the carrier and foundation of others. If ecological security cannot be effectively ensured, homeland, economic and social security will certainly be affected, thus threatening regional security, national security and even global security.

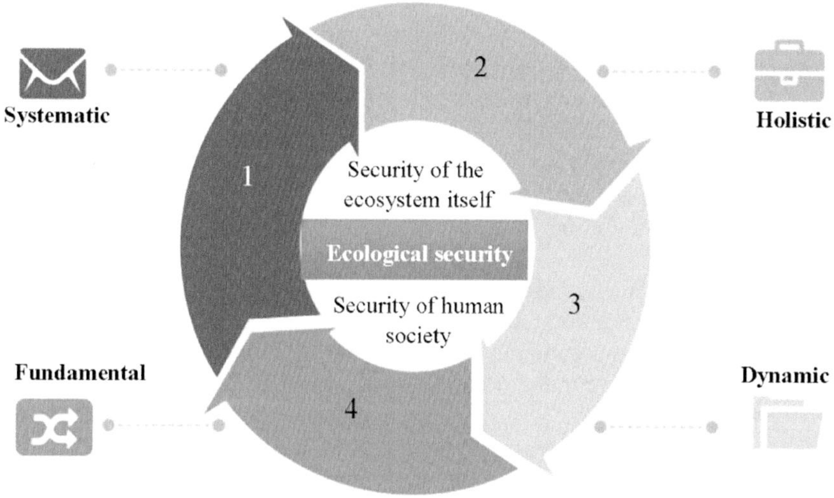

Characteristics of ecological security

1.2 Progress of China's Ecological Security

Since ancient times, China has been prone to natural disasters, so traditional Chinese culture always attaches importance to the ecological environment, emphasizing that "the benevolent love mountains, while the wise love water", and promoting the ideas of "Dao follows the laws of nature" and "man is an integral part of nature". However, in modern times (especially after the First Opium War in 1840, also known as the Anglo-Chinese War), China's ecological environment was seriously damaged due to long-term foreign invasion and internal warfare, and after the People's Republic of China was founded in 1949, the Chinese government and its people, who deeply suffered from the damaged eco-environment, began to make efforts to gradually improve environmental protection and ecological security.

In the first period (1949–1978), China began to explore ways to ensure ecological security. After the People's Republic of China was founded in 1949, to address the problems of dirty, unhygienic environments and serious epidemics of infectious diseases, the Chinese government launched a massive patriotic health campaign nationwide. In just six months, more than 15 million tons of garbage were removed, 280,000 kilometers of ditches were dredged, 1.3 million new and existing wells were dug and redug, 4.9 million new and existing toilets were built and rebuilt, a large number of sewage ponds were filled, and a great number of rats and flies were eliminated, which greatly improved the rural and urban environments. In 1955, faced with devastated and bare land, the Chinese government called for "greening the motherland" and "greening all the barren land and mountains that could be greened", setting off a wave of planting forests, greening the motherland and reconstructing the country. In 1972, China sent a delegation to participate in the United Nations Conference on Human Environment held in Stockholm, Sweden. In 1973, the Chinese government held the first National Environmental Protection Conference, which formally incorporated environmental protection into the functions of governments at all levels. In 1978, the will of the state to "protect the environment and natural resources, and prevent pollution and other public hazards" was embodied in the *Constitution of the People's Republic of China* for the first time, and China began to gradually promote ecological security gradually.

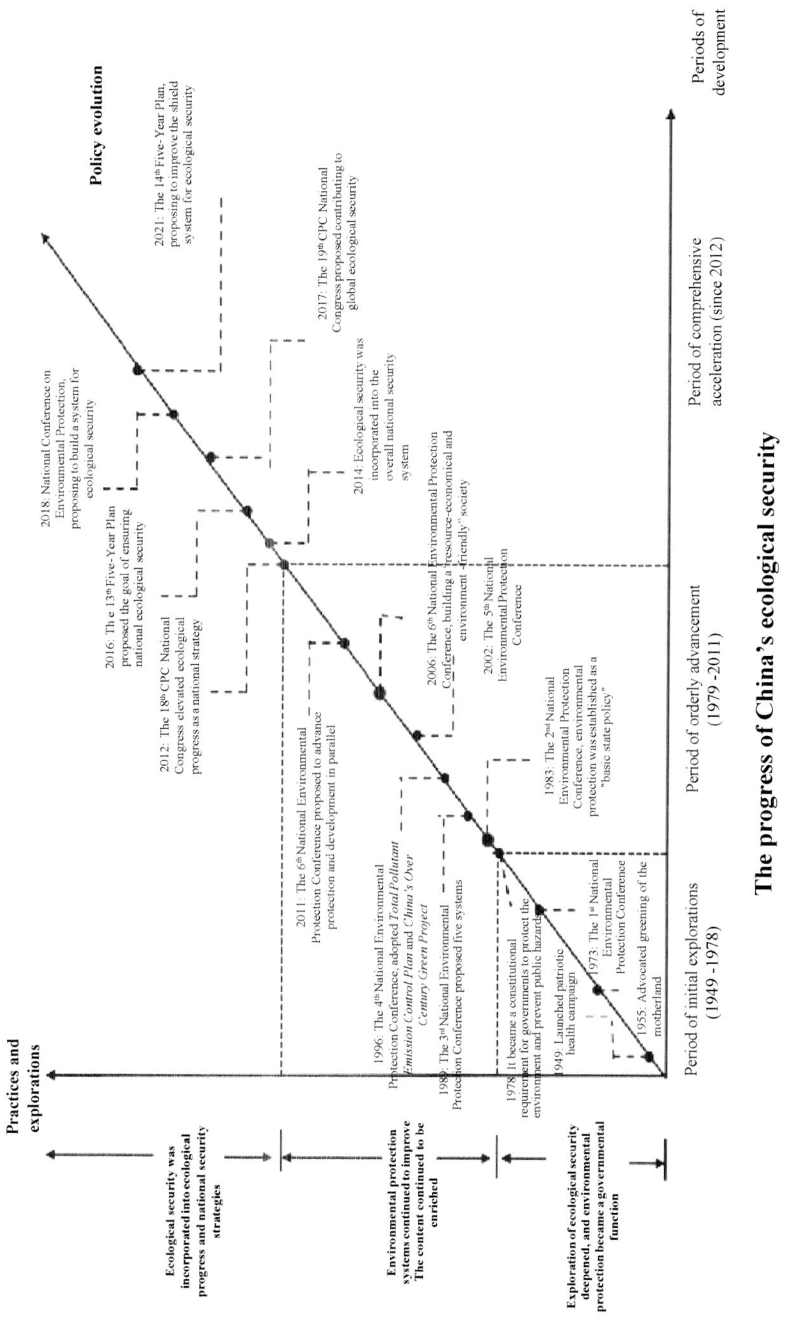
Policy evolution
Practices and explorations
Periods of development
2021: The 14th Five-Year Plan, proposing to improve the shield system for ecological security
2018: National Conference on Environmental Protection, proposing to build a system for ecological security
2017: The 19th CPC National Congress proposed contributing to global ecological security
2016: The 13th Five-Year Plan proposed the goal of ensuring national ecological security
2014: Ecological security was incorporated into the overall national security system
2012: The 18th CPC National Congress elevated ecological progress as a national strategy
2011: The 6th National Environmental Protection Conference proposed to advance protection and development in parallel
2006: The 6th National Environmental Protection Conference, building a "resource-economical and environment-friendly" society
2002: The 5th National Environmental Protection Conference
1996: The 4th National Environmental Protection Conference, adopted Total Pollutant Emission Control Plan and China's Over Century Green Project
1989: The 3rd National Environmental Protection Conference proposed five systems
1983: The 2nd National Environmental Protection Conference, environmental protection was established as a "basic state policy"
1978: It became a constitutional requirement for governments to protect the environment and prevent public hazards
1973: The 1st National Environmental Protection Conference
1955: Advocated greening of the motherland
1949: Launched patriotic health campaign
Period of initial explorations (1949-1978)
Period of orderly advancement (1979-2011)
Period of comprehensive acceleration (since 2012)
Ecological security was incorporated into ecological progress and national security strategies
Environmental protection systems continued to improve The content continued to be enriched
Exploration of ecological security deepened, and environmental protection became a governmental function
The progress of China's ecological security

In the second period (1979–2011), China boosted ecological security in an orderly manner. After China launched the reform and opening-up in 1978, the country made great achievements in economic and social development, but the situation of ecological security became increasingly serious due to environmental pollution, vegetation deterioration, land desertification, biodiversity decrease and ecosystem dysfunction. In 1983, China held the second National Environmental Protection Conference, which formally established environmental protection as a "basic state policy". During the period between 1979 and 2011, China held six national conferences on environmental protection, continuously improved the policies and systems for ecological protection, and promoted national ecological security by formulating and implementing five-year plans one after another. For example, during the seventh Five-Year Plan period, China issued the first five-year environmental plan, the *Seventh Five-year Plan for National Environmental Protection*, and in the eighth Five-Year Plan period, China formulated the *Ten Countermeasures for Environmental and Development Problems*. In the ninth Five-Year Plan period, China introduced the *Total Pollutant Emission Control Plan for the Ninth Five-Year Plan Period* and *China's Over Century Green Project*. During the 10th Five-Year Plan period, China promoted the transformation from end-of-pipe treatment to regulation over operation. During the 11th Five-Year Plan period, China proposed building a "resource-economical and environmentally friendly society" and strengthening such social construction as an important way to ensure ecological security.

Now China is in the third period, which started in 2012. After the 18th CPC National Congress that was held in 2012, ecological security was gradually elevated to a national security strategy. In 2014, the National Security Commission of the Communist Party of China held the first meeting and put forwards a "holistic approach to national security" and formally incorporated ecological security into the overall national security system. In 2016, the 13th Five-Year Plan set the development goal of "shaping the overall national ecological security pattern and safeguarding national ecological security." In 2017, the report of the 19th CPC National Congress proposed advancing ecological security in a scientific and balanced way. In 2018, China held the National Conference on Environmental Protection to accelerate the construction of an ecological security system that emphasizes the maintenance of well-functioning ecosystems and the effective prevention and control of environmental risks. Improving the ecological security system and building a solid national barrier for ecological security have become important targets for China to promote ecological progress and national security in the new era.

2 Drawing Strict "Redlines for Ecological Conservation"

As the contradiction between economic and social development and resources and the environment becomes sharper, to secure sustainable development, China faces many major challenges, including tightening resource and environmental constraints due to serious environmental pollution and ecosystem degradation, which form a severe threat to the territorial ecological security pattern in China. Against this background, China has been exploring innovations in the development of a national ecological security pattern by strictly implementing the "red lines for environmental protection" and building a sound barrier for ecological security, striving to ensure national and regional ecological security.

2.1 Drawing Ecological Conversation Red Lines Scientifically

The ecological conservation red lines are the bottom lines for natural and ecological security, the quantitative bottom lines for ecological land and species to safeguard and maintain the security of territorial ecology, human habitats and biodiversity, the bottom line and lifeline to ensure and maintain national ecological security, and the most important ecological spaces. China has spatially planned to draw ecological conservation red lines to cover ecologically vulnerable areas and important ecologically functional zones, such as zones with important functions of water conservation, biodiversity maintenance, soil and water conservation, wind and sand control. In 2018, the National Conference on Environmental Protection stressed the need to accelerate the delineation and strictly adhere to the "red lines for environmental protection, minimum standards for environmental quality, and ceilings on resource utilization". Subsequently, various local governments developed an ecological and environmental access list, and governments at all levels are required not to get around or lower the standards in local legislation, policy formulation, planning, enforcement and supervision.

In recent years, China issued the *National Ecological Conservation Red Lines: Technical Guide for the Delineation of Ecological Function Baselines (for Trial Implementation)*, *Technical Guide for the Delineation of Ecological Conservation Red Lines*, *Notices on Piloting Ecological Conservation Red Lines*, *Some Opinions on Drawing and Strictly Adhering to Ecological Conservation Red Lines*, and *Guides for Drawing Ecological Conservation Red Lines* to make ecological redlines rigid constraints to promote the protection and restoration of ecosystems, demarcate the boundaries for ecological redlines, specify the primary responsibilities at all levels, and strengthen monitoring and supervision for ecological conservation and restoration. At the same time, it requires firmly adhering to the basic requirements that pollutant emissions must not exceed the maximum carrying capacity of the environment and that environmental quality must not worsen and may only improve to effectively prevent and control ecological and environmental risks. By 2021, China had initially established an eco-environment zoning regulatory system centering on the "three lines and one list" of the red lines for environmental protection, minimum standards for environmental quality, ceilings on resource utilization and eco-environment access list, which basically covers all ecological spaces and strictly safeguards the boundaries for natural ecological security in all aspects.

2.2 *Building a Solid Barrier for Ecological Security*

To improve the development of territorial space, promote the coordination among population, economy, resources and the environment, and ensure the ecological security of territorial space, China formulated and promulgated the *National Plan for Developing Priority Zones* in 2011, which was based on the strategic land and sea ecological security layout and highlighted the ecological support to major national strategies for social and economic development, aiming at building a land ecological barrier composed of Qinghai-Xizang Plateau ecological shield, the Loess Plateau-Sichuan-Yunnan ecological shield, the northeast China forest belt, the northern China desertification-prevention belt, and the southern China mountainous belt, as well as an ecological security pattern that takes the important water system of large rivers as the framework and that is supported by other key ecological function zones and mainly composed of areas prohibited development. In 2021, the 14th Five-Year Plan further proposed to "improve the system of shields for ecological security", and on the basis of existing shields, accelerate the construction of ecological barriers such as the Yellow River key ecological function zone, the Yangtze River key ecological function zone, Qinghai-Xizang Plateau ecological shield (zone), the northeast China forest belt, the northern China desertification-prevention belt, the southern China mountainous belt, and the coastal belt to strengthen the ecological conservation and treatment of major rivers and important lakes and wetlands and to promote the establishment and protection of ecological corridors, in hope of forming a pattern of "three zones and four belts" and solid barriers for ecological security and safeguarding the boundaries of ecological security.

3 Intensifying the Efforts in Eco-Environmental Conservation and Restoration

The natural ecosystems in China face serious problems of degradation and loss of functions, so strengthening ecological conservation and restoration of severely damaged ecosystems has become an important part of China's ecological construction in recent years. Under the guidance of the philosophy of ecological civilization, China upholds the basic policy of giving priority to environmental protection and letting nature restore itself, insists on respecting, accommodating to and protecting nature, and follows the laws of natural ecosystems, giving full play to nature's ability to repair itself and striving to maintain a virtuous cycle of ecosystems.

3.1 Promoting the Protection and Restoration of Key Ecosystems

The implementation of major projects for the protection and restoration of key ecosystems is one of the important measures to promote ecological security in China. In June 2020, the China National Development and Reform Commission and the Ministry of Natural Resources jointly issued the *Master Plan for the Major Projects for the Protection and Restoration of National Key Ecosystems (2021–2035)*, which set out the development goals, namely, to achieve fundamental improvement in the condition of natural ecosystems such as forests, grasslands, deserts, rivers, lakes, wetlands and oceans, significantly enhance the stability of ecosystems, and basically realize a virtuous cycle of natural ecosystems and the harmonious coexistence between humans and nature.

While building national ecological barriers, China focuses on improving the ecological protection and restoration of national key ecological function areas, areas covered by ecological conservation red lines, key national nature reserves and other areas and makes every effort to solve the core ecological problems in a number of key areas. Relying mainly on the Qinghai-Xizang Plateau ecological shield zone, the Yellow River key ecological zone, the Yangtze River key ecological zone, the northeast China forest belt, the northern China desertification prevention belt, the southern China mountainous belt, and the coastal belt, China is holistically considering the integrity of ecosystems, the continuity of geographical units and the sustainability of economic and social development and actively advancing a series of major projects for ecosystem protection and restoration, including natural forests protection projects, protection and restoration projects for grasslands, rivers, lakes, wetlands and marine ecosystems, biodiversity protection projects, invasive alien species prevention and control projects, shelter forest programs, programs returning marginal farmland into forests and grasslands, and programs returning polder lands to lakes and wetlands. In addition, China is taking comprehensive steps to control over-extraction of groundwater, desertification, rocky desertification and soil erosion, to restore the ecology in mining areas, and to carry out the Blue Bay environmental improvement initiative, which plays an important role in curbing the deterioration of various natural ecosystems and lays a solid foundation for the virtuous cycle of ecosystems.

For example, to protect the ecosystem of the Yangtze River Basin, the Chinese government has formulated and implemented the *Environmental Protection Plan for the Yangtze River Economic Belt*, promulgated the *Yangtze River Protection Law of the People's Republic of China*, and issued a series of protection programs and measures, including the *Opinions of the General Office of the State Council on Strengthening the Protection of Aquatic Life in the Yangtze River*, the *Implementation Plan for Further Promoting Prohibiting Fishing in Key Waters of the Yangtze River and Ensuring the Livelihoods of Affected Fishermen*, and the *Special Action Plan Against Illegal Fishing in the Yangtze River Basin*. In recent years, China upheld the philosophy of "mountains, rivers, forests, fields and lakes are part of a community of life", focused on the integrated protection of water resources, ecology and environment, and intensified the rigid constraints of the ecological environment to ensure that the eco-environmental quality of the Yangtze River would only get better and never become worse. Poyang Lake is an important lake in the Yangtze River

Basin and the largest freshwater lake in China. With these measures to improve water quality, regulate fishing and extraction of sediment and gravel, prevent and control invasion of foreign species, maintain the cycle of the freshwater system, and restore and protect the aquatic habitat, the lake has now become a paradise for wild animals such as white cranes and finless porpoises.

Poyang Lake priority area for biodiversity conservation

Source "The Candidate of Outstanding Environmental Planning, Poyang Lake Nature Reserve", http://gongyi.sina.com.cn/greenlife/2011-05-09/155126208.html (the picture above). "Finless

Porpoises Appear in Poyang Lake in Groups, Demonstrating the Good Ecology in Jiangxi", https://m.gmw.cn/baijia/2020-05/08/1301209994.html (the picture below)

3.2 *Promoting the Restoration of Damaged Ecosystems*

Due to long-term human activities in the past, in particular modern and contemporary wars, and high-intensity development and construction since 1949, especially after the reform and opening-up, the ecosystems in China have been severely damaged and degraded. In response to the tough problems of natural ecosystems, China has continuously implemented a series of targeted ecosystem restoration projects, such as afforestation and land desertification control programs and systematic restoration and comprehensive treatment projects for mountains, rivers, forests, fields, lakes, grasslands and deserts as a community of life, creating many models of ecological restoration for the world.

Taking afforestation as an example, shortly after the founding of the People's Republic of China, the Chinese government launched a nationwide campaign to plant trees and green the motherland. As Chinese leaders set good examples, the people had high enthusiasm, so the campaign lasted for a long time. According to the eighth National Forest Resources Survey (2009–2013), compared with the first Forest Resources Survey (1973–1976), the area of forests in China increased by 90 million hectares, the forest coverage rate rose by 8.9 percentage points, and the forest stock volume grew by 6.48 billion cubic meters. From 2013 to 2020, the forest coverage rate in China further rose by 2.68 percentage points to 23.04%, and the forest stock volume had a net increase of 3.839 billion cubic meters. Since 2004, China realized double decreases in the area of land becoming deserts and rocky deserts for three consecutive monitoring periods. As a paper published in the British journal *Nature Sustainability* in February 2019 pointed out, afforestation in China contributed approximately a quarter of new global afforested areas between 2000 and 2017. In the case of Saihanba Jixie Forestry Farm in Hebei Province, which is adjacent to Beijing, Tianjin and the Inner Mongolia Autonomous Region, the farm was once rich in water resources and covered by dense forest and grasses and an important part of the royal hunting grounds, "Mulan Paddock", in the Qing Dynasty. However, due to overcultivation and grazing, together with wars and wild fires, Saihanba turned into a wasteland with frequent sandstorms and little vegetation, and the sandstorms even affected Beijing, Tianjin, Hebei Province and other areas at far distances. After decades of afforestation and soil erosion control, especially ecological restoration in the last 10 years or so, Saihanba has now been covered by forests again, and the wasteland with "yellow sandstorms sheltering the sky and no trees for birds to rest" has gone, turning into a green ecological barrier and safeguarding the water sources for the capital and northern China. In 2017, the builders of Saihanba Jixie Forestry Farm were awarded "Champions of the Earth" by the United Nations Environment

Programme, and in 2021, they again won the "Land for Life Award", the highest honour of the United Nations in the field of combating desertification.

Before (the picture above) and after (the picture below) Saihanba Jixie Forestry Farm's afforestation

Source Qiao Ya, Gao Hang, "Saihanba Jixie Forestry Farm Won the 'Land for Life Award', the Highest Honor of the United Nations in the Field of Combating Desertification", http://report.hebei.com.cn/system/2021/09/29/100781888.shtml?from=singlemessage

4 Innovations in Biodiversity Protection

Biodiversity is the most important part of the life support system, with ecological functions such as protecting the environment and mitigating natural disasters, while playing an important role in maintaining the virtuous cycle of ecosystems and ecological balance and providing the majority of food for humans. It is the essential foundation for human survival and sustainable development. However, due to disturbing, invading and damaging human activities, global biological habitats are quickly shrinking in size and dysfunction, leading to sharp decreases in the number of many species and even extinction. In other words, global biodiversity loss poses a major risk to ecological security on Earth and sustainable human development. According to the *Global Assessment Report on Biodiversity and Ecosystem Services* issued by the United Nations Environment Programme (UNEP) in 2019, 66% of marine areas were under increasing cumulative impacts, and 85% of wetlands disappeared. On average, approximately one-quarter of species in 21 countries with detailed assessment records were threatened in the past five decades, meaning that approximately one million species are already endangered and species worldwide are becoming extinct at a rate at least tens to hundreds of times higher than the average of the last 10 million years. According to the *Nature Risk Rise* released by the World Economic Forum (WEF) in June 2020, the global human population of 7.6 billion accounts for only 0.01% of the total biomass on Earth but has caused the extinction of 83% of wild mammals and 50% of plants on this planet, meaning that global ecological security is in a critically serious situation. Vast land territories in China give birth to all types of terrestrial ecosystems on earth, making China is one of the most biodiverse countries in the world. Biodiversity conservation in China is of great significance to global ecological security.

4.1 Biological Security is Included in the National Security System

China was one of the first countries to sign and approve the *United Nations Convention on Biological Diversity* (CBD), and on September 30th, 2020, China called on the people of all countries and regions to "work together to build a beautiful world in which all live in harmony!" At the leaders' summit of the 15th meeting of the Conference of the Parties to the *Convention on Biological Diversity* (COP15) held in Kunming, China, in October 2021, China actively advocated building a community of all life on Earth, proposing to build a global home in which humans and nature coexist in harmony, the economy and the environment advance coordinately, and all countries/regions around the world develop sustainably.

In recent years, China has strengthened biodiversity conservation in all aspects and formulated and promulgated *the Biosecurity Law of the People's Public of China, Wild Animal Protection Law of the People's Public of China, Seed Law of the People's Public of China* and other laws and regulations. Meanwhile, China established the National Committee for Biodiversity Conservation (CNCBC), which is composed of departments under the State Council, and carried out a series of action plans, such as the *China National Biodiversity Conservation Strategy and Action Plan (2011–2030)* and the *Opinions on Further Strengthening Biodiversity Conservation*, meaning that biosecurity has become a national strategy and is incorporated into the medium- and long-term plans of various sectors and regions in China. To better preserve biodiversity, China has implemented continuous biodiversity background surveys and observations, carried out rescue protection of endangered wildlife, promoted the protection of biogenetic resources, and enhanced the regulation on wildlife import and export to prevent biosecurity risks. China has formulated and put into force the *Master Plan for the Major Projects for the Protection and Restoration of National Key Ecosystems (2021–2035)* and promoted the construction of major projects for nature reserves and wildlife protection established a nature reserve system based on national parks and released the *China Biodiversity Red List* to effectively preserve biodiversity in key areas and form a more complete species protection system. In addition, the Chinese government has strengthened law enforcement and supervision for biodiversity conservation to combat illegal and criminal activities that damage biodiversity, such as actively conducting collaborated trans-department, cross-region and cross-border joint actions, central ecological environmental protection inspections, and special law enforcement campaigns such as Green Shield inspections of nature reserves.

4.2 Ensuring the Safety of Biological Habitats

Habitat refers to the space for a species to live in, and preserving the habitat means protecting the Earth, the common homeland. Since 2012, China has implemented major projects for the protection and restoration of important ecosystems, continuously improved and optimized the shield system for ecological security, carried out a series of biological habitat protection actions, established various nature reserves with national parks as a main type, effectively strengthened the protection and management of nature reserves such as Sanjiangyuan National Park, Qilian Mountain State Nature Reserve, Northeast China Tiger and Leopard National Park, Giant Panda National Park, Hainan Tropical Rainforest National Park and Qomolangma National Nature Reserve, enhanced the protection and restoration of important natural ecosystems, natural relics and natural landscapes, and built an overall protection network for important native ecosystems.

In addition, the Chinese government is actively promoting the construction of a nature reserve system of national parks, nature reserves, nature parks and so on and has designated 32 priority areas for biodiversity conservation on land and three marine areas, accounting for approximately 29% of China's land mass. China has established 749 observation plots nationwide for birds, amphibians, mammals and butterflies, constituting the China Biodiversity Observation Network (China BON), with a total of 11,887 transects and sampling points, so that 90% of vegetation and terrestrial ecosystem types, and 85% of key wild animal populations are under effective protection.

First, China attaches great importance to improvements in in situ conservation systems and the construction of national parks. In situ conservation means protecting living beings and the original ecological environment, which can preserve the original characteristics of organisms to the greatest extent, and it is also an important method and measure for biodiversity conservation. China has established an in situ nature reserve system based on national parks and nature reserves, supplemented by various types of nature parks, which account for approximately 18% of its land mass. Since 2015, China has set up 10 pilot national parks, including Sanjiangyuan National Park, accounting for approximately 2.3% of the land mass, putting 90% of terrestrial ecosystem types and 71% of key wildlife species under effective protection.

Located in the hinterland of the Qinghai-Xizang Plateau, Sanjiangyuan National Park is the source of three major rivers in China, the Yangtze, Yellow and Lancang rivers, making it an important freshwater source in both China and Asia, and there are glaciers and snowy mountains, high-altitude wetlands, gobi deserts, alpine grassland and meadows. With a protected area of 190,700 square kilometers, it provides overall protection for the sources of the Yangtze, Yellow and Lancang rivers, and it is also a protection model of the large-scale alpine ecosystem on the Qinghai-Xizang Plateau, the third pole of the Earth.

The Northeast China Tiger and Leopard National Park in Jilin and Heilongjiang provinces in northeast China, bordering Russia and the Democratic People's Republic of Korea, is home to the largest and only sustainable wild Siberian tiger and Siberian leopard population in China. With a protected area of 14,100 square kilometers, there are also typical temperate forest ecosystems in the park. The park now adopts scientific and comprehensive measures for protection, ecological restoration, species conservation and rescue, and international cooperation and has become a model of cross-border cooperative conservation.

Scenes of Sanjiangyuan National Park

Source “The Father of Mountains and Source of Rivers in Qinghai Province”, https://www.sohu.com/a/432202507_732951. (the picture above). Qinghai Daily, “The Scenery in Kekexili”, http://sjy.qinghai.gov.cn/park/dzwbh/24390.html (the picture below)

Second, China pays special attention to improving the ex situ biological conservation system and makes every effort to protect biological species. Ex situ conservation means relocating the species out of the original locations and into zoos, botanical gardens, aquariums and endangered animal breeding centers for scientific protection, and it is an important means of protecting species when the conditions are suitable for their survival no longer exist, when it is difficult for the environment to sustain their survival and evolution, or for these critically endangered species, which hardly sustain the population naturally. China has made efforts to improve the ex situ biological conservation system and to demonstrate the scientific spirit in biological rescue and breeding.

Northeast China Tiger and Leopard National Park

Source Jilin Channel at People's Daily app, "The Northeast China Tiger and Leopard National Park: the First National Parks in China, and Home to Approximately 50 Wild Siberian Tigers", http://www.jllydj.gov.cn/szxw/20777.jhtml

China has constructed nearly 200 botanical gardens and preserved more than 23,000 species of plants while establishing 250 wildlife rescue and breeding bases, germplasm resource banks and gene banks, helping more than 60 rare and endangered wild animal species through artificial propagation. Now, China has set up a relatively complete ex situ conservation system.

4.3 Launching Well-Planned Campaigns to Save Endangered Species

The extinction of one species will destroy the integrity of the biological chain, causing chain reactions affecting the existence of other species, breaking the balance of the ecosystem, and leading to many ecological security problems. Rescuing endangered species is conducive to preserving rare and valuable genes and biodiversity and of great significance to sustainable human development. Well-planned campaigns to save endangered species in China have made wonderful successes in biological rescue and protection. In 2021, 980 species and eight categories of wild animals were included in the *List of Key Protected Wild Animals* in China, of which giant pandas, black crested gibbons, Przewalski's gazelles, brown eared pheasants, Yangtze finless porpoises, Dabry's sturgeons and Chinese alligators are endemic to China, and 455 species and 40 wild plants were included in the *List of Key Protected Wild Plants in China*, including species endemic to China such as Abiesbeshanzuensis, Metasequoiaglyptostroboides, Dendrobiumhuoshanense and Aquilariayunnanensis. China has carried out rescue protection plans for some rare and endangered wild animals and for 48 critically endangered wild animals and their habitats, including giant pandas, Asian elephants, black crested gibbons, and Siberian tigers.

The captive population of breeding pandas has grown in number and quality. Over the past 40 years, the population of giant pandas in the wild has grown from 1,114 to 1,864. Giant pandas have been downgraded from "endangered" to "vulnerable" on the list of species at risk of extinction, and some artificially bred giant pandas have been released into natural habitats and integrated into the wild population successfully.

Giant pandas

Source sohu.com, "Closely Encountering the Giant Panda Babies All Day Long, as Assistants of Scientists and Volunteers in Field Scientific Research! Take Action in This Summer Vacation to Protect the Earth!", https://www.sohu.com/ a/243597923_658489

4.4 Promoting Animal-Friendly Campaign

Both humans and animals are important parts of ecosystems, and animals are integral to ecological balance. Humans and animals are interdependent. With the development of human society, the human population keeps growing, while the animal population and the number of species decrease rapidly, and some even become extinct, so it is of important significance to protect animals and encourage animal-friendly activities for human survival. China continues to promote publicity and education on caring for the environment and protecting animals and plants and advertises the idea of building a beautiful homeland in which humans and nature can coexist in harmony while conducting special campaigns to combat illegal and criminal activities concerning wildlife. For example, as Tibetan antelopes in the Hohxil region in the Qinghai-Tibet Plateau begin a great migration every spring and summer for breeding, the Qinghai-Xizang Railway, from design to construction and then to operation, follows the principle of reducing the impact on the surrounding eco-environment to the minimum, and special passageway bridges were added for Tibetan antelopes and other wild animals. In addition, workers often stopped waiting patiently for Tibetan antelopes to pass when constructing the railroad. All these examples of wild animal protection fully demonstrated the biodiversity conservation practices in major construction projects in China.

Harmonious coexistence of Qinghai-Xizang Railway and wild animals

Source CRI, "The Results of the first Asia Press Photo Contest Announced and the Selected Collection of Award-winning Works (a group of photos)", http://news.cctv.com/society/20070602/101448.shtml (the picture above). Xinhua News Agency, "The Qinghai-Xizang Railway Has Special Passageways for Wild Animals", http://news.sohu.com/20060628/n244014021.shtml (the picture below)

In March 2020, 15 wild Asian elephants left the Xishuangbanna Nature Reserve and moved all the way north. In this migration of 17 months, these Asian elephants headed northwards and roamed 1,300 km across half of Yunnan province before returning, attracting the attention not only of Chinese people in great numbers but also various international media. Yunnan provincial authorities deployed police officers and workers 25,000 person-times, 937 sorties of drones and 150,000 emergency vehicles to safeguard migration. During the migration of the elephant herd, no harm was done to the elephants, and they all returned home unharmed. The picture of harmony between humans and elephants moved the world.

Migrating wild Asian elephants in Yunnan Province, China

Source Han Xianyang, Zhang Guannan, Fu Zhigang, Zhang Renzhong, and Wang Sicheng, "The Migration of the Elephants Demonstrates the Harmony between Humans and Nature", *Guangming Daily*, June 22, 2021, p.12

5 Winning a Determined Battle to Prevent and Control Pollution

For a long time in the past, water, soil and marine pollution, as well as pollution of agricultural products, threatened people's physical and mental health and sustainable development, particularly heavy air pollution, black and odorous water bodies, garbage surrounding cities and rural environmental problems, which seriously affected people's production and lives. In 2011, China issued the *Opinions of the State Council on Promoting Major Tasks for Environmental Protection*, marking "environmental risk prevention" as a written requirement of the State Council, and China focused on solving outstanding ecological and environmental problems and improving the living environment. Subsequent improvements in eco-environmental quality have helped ecological habitats and various ecosystems restore themselves and their functions, effectively preventing and controlling the risks of ecosystem degradation and biodiversity loss. Since environmental pollution prevention and control is a long-term and arduous task, in the 14th Five-Year Plan period and beyond, China will continue to wage the determined battle against pollution, implement positive and effective moves toward the targets of carbon peaking and carbon neutrality, and vigorously promote synergizing the reduction of pollution and carbon emissions to provide the people with a safe and healthy ecological environment and lay a good ecological foundation for China to embark on a new journey to build a modern socialist country in all respects.

5.1 Winning a Decisive Victory in the Campaign to Make Skies Blue Again

Emissions from human production and living activities of particulate matter (PM10), fine particulate matter (PM2.5), SO2, CO2 and other gaseous substances have altered the composition of the air, resulting in damage to the health of humans, plants and animals and even causing severe air pollution that seriously affects production and living activities. China has always attached great importance to the prevention and control of air pollution, since it promulgated the *Law on Air Pollution Prevention and Control of the People's Republic of China*, continued to implement the *Action Plan for Prevention and Control of Air Pollution*, developed a "1 + N" policy framework for carbon peak and carbon neutrality, and issued the *Guiding Opinions on Strengthening Source Prevention and Control of High Energy Consuming and High Emission Projects*. Meanwhile, China continues to promote the coordinated control of (PM2.5) and ozone (O3) emissions and comprehensive prevention and control of air pollution by "dynamically" eliminating these "scattered, noncompliant, and polluting" enterprises and freight vehicles with excessive emissions, fully implementing China VI emission standards for light-duty vehicles, advancing comprehensive regulations on open-pit mines and treatment of ground dust, and boosting the optimization and adjustment of industrial, energy, transportation and land-use structure. As a result, China has effectively improved air quality and ensured atmospheric environmental security.

In 2021, the average PM2.5 concentration in Chinese cities at or above the prefecture level was 30 $\mu g/m^3$, and the average PM10 concentration was 54 $\mu g/m^3$, down 40.0% and 37.9%, respectively, from 2015; the proportion of days with good air quality in cities reached 87.5%, up 10.8% points from 2015; the number of days with blue skies in all regions increased significantly, air quality continued to improve, and the number of days with heavily polluted weather decreased significantly. Beijing has focused on PM2.5 pollution since 2016, vigorously promoted clean energy strategies, and eliminated 1.125 million high-emission vehicles promoted the sale of new energy vehicles (485,000 units) and reduced carbon dioxide emissions per USD 1,384 GDP to only 0.41 tons, bringing the proportion of days with good air quality in Beijing up by 24.8% points in 2021 compared to 2016. The experience of Beijing was included in the United Nations Environment Programme's practical cases for air pollution control.

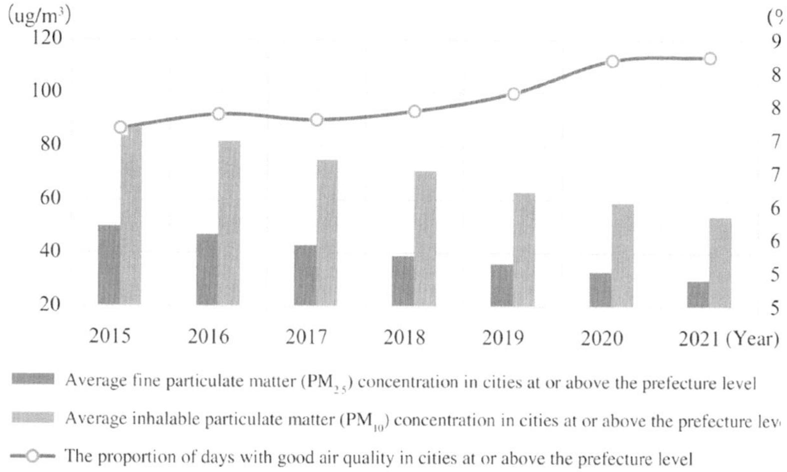

Improvements in major air quality indexes in cities in China between 2015 and 2021

Drawn based on *Reports on the State of the Ecology and Environment in China (2015–2020)* and the *Report of the State Council on the environmental state and the accomplishment of environmental protection targets in 2021*

5.2 *Concentrating Strengths on the Campaign to Make Water Clear Again*

Industrial wastewater, domestic sewage, and pollutants from agricultural irrigation and the use of pesticides and chemical fertilizers cause deterioration of water quality and death of aquatic life, endanger human health, and damage water ecosystems, seriously threatening human production and sustainable development. China attaches great importance to water pollution prevention and control, as it promulgated the *Law of the People's Republic of China on Water Pollution Prevention and Control* and continuously implemented the *Action Plan for Prevention and Control of Water Pollution* to jointly promote pollution reduction and water ecological capacity expansion. Specifically, China carried out in-depth standardization of centralized drinking water sources, concentrated efforts to treat black and odorous water bodies in urban and rural areas, strengthened the control of rural domestic sewage, and promoted comprehensive environmental improvement in rural areas, conducted investigations of sewage outfalls along major rivers, improved eco-environmental protection of important watersheds, coasts and groundwater, and carried out the Blue Sea initiative, effectively preventing water pollution and ensuring water ecological security.

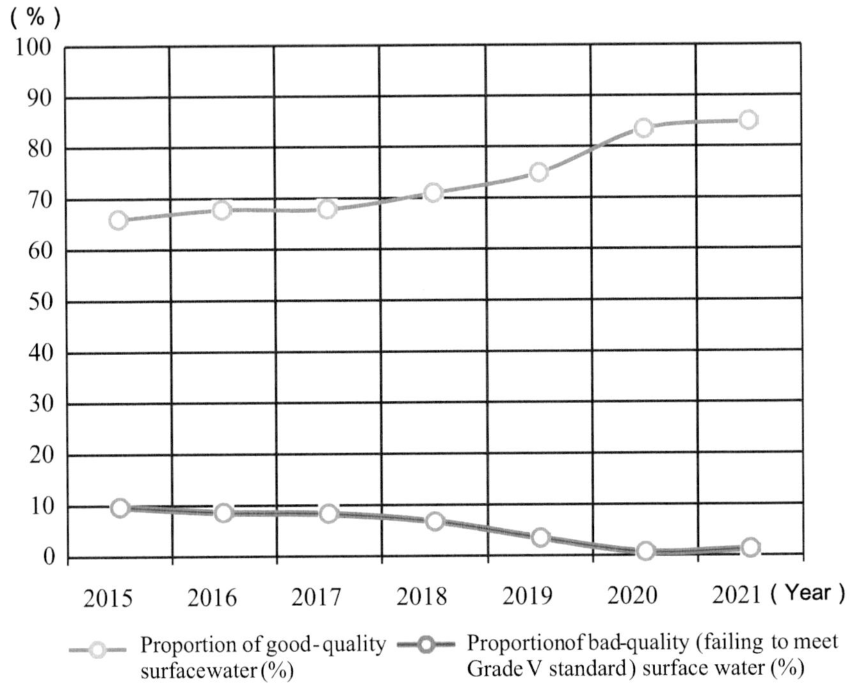

Changes in major indexes for national water environmental quality (2015–2021)

Drawn based on *Reports on the State of the Ecology and Environment in China (2015–2020)*, and the *Report of the State Council on environmental state and the accomplishment of environment protection targets in 2021*

In 2021, the proportion of good-quality surface water bodies in China reached 83.5%, and the proportion of river basins with good water quality reached 87.4%, up by 19 and 15.3% points from 2015, respectively. In addition, 98.2% of black and odorous water bodies in built-up areas of cities at or above the prefecture level nationwide disappeared, and the proportion of good-quality marine waters reaching Grade I was 97.7% in terms of area, indicating that the proportion of good-quality water ecology rose continuously.

Taihu Lake after ecological treatment

Source "After A Decade of Treatment, the Water Area of Taihu Lake in Wuxi Recovered to the Quality Before 1997", https://jnews.xhby.net/waparticles/10/qlhBUagZRB9ONBBO/1; "Suzhou Municipal Government Reuses 'Taihu' in Naming Administrative Zones, A Wish for long of Suzhou Residents", https://www.163.com/dy/article/DJFPMD8T0524VF0A.html

Taking Taihu Lake as an example, which is one of five largest freshwater lakes in China, in 2007, there was a blue algal bloom in Taihu Lake, and the water turned black and odorous, resulting in a local water supply crisis and the death of aquatic plants and animals. In the following 12 years, local governments carried out continuous treatment of Taihu Lake, covering the water and the shore, and implemented projects and measures including urban and rural sewage treatment, industrial and agricultural pollution control at the sources, integrated small watershed management and ecological restoration, as well as monitoring and early warning of algal blooms in the lake, restoring the water in the whole lake and each lake area to a light eutrophic state, and achieving the governance goal of "ensuring the safety of drinking water and preventing algal bloom in large area". Now, Taihu Lake has become a beautiful scenic spot again, with clear water, birds and flowers.

5.3 Solidly Deploying the Battle Against Soil Pollution

Soil pollution directly or indirectly contaminates various human foods, such as grains, vegetables, fruits and livestock products, causing food safety problems and endangering human health. China attaches great importance to soil pollution prevention and control and has fully launched agricultural source pollution control since 2015, as well as action plans to reduce the use of chemical fertilizers and pesticides. In 2017, China achieved the goal of zero growth in the use of pesticides and chemical fertilizers three years ahead of schedule. In 2016, China began the continuous implementation of the *Action Plan for Soil Pollution Prevention and Control*, strengthened the pilot application of soil pollution treatment and restoration technologies and the construction of pilot zones on soil pollution comprehensive prevention and control, promoted "waste-free cities", conducted special investigation and regulation campaigns for hazardous waste environmental risks, and advanced the source prevention and control of agricultural pollution and reuse of wastes as useful resources to ensure the safety of agricultural products. By 2020, the use of pesticides and fertilizers in China was reduced by 420,000 tons and 7.72 million tons respectively, compared with the peaks, the safe utilization rate of contaminated arable land nationwide reached 90% or so, and that of contaminated land topped 93%, successfully completing the treatment targets set in the *Action Plan for Soil Pollution Prevention and Control,* effectively ensuring food and soil edaphon security, basically putting soil environmental risks under control and curbing the trend of worsening soil pollution.

China comprehensively promoted environmental protection and ecological conservation, not only significantly improving its ecological security but also contributing to global ecological security. A survey conducted by the National Bureau of Statistics in 2020 showed that the Chinese public had a degree of satisfaction with the ecological environment as high as 89.5%, up by 10.7% points over 2017, and people's satisfaction and sense of gain with the ecological environment has also significantly improved.

Chapter 5
Optimizing the Layout of Territorial Space

Jingjing Shan, Guangyao Sheng, Wei Lou, and Qiang Li

The territory is the largest public resource of a country and region. The vast territorial space of land and sea constitutes the homeland of 1.4 billion Chinese people seeking common prosperity and sustainable development and is also a spatial carrier and an important part of ecological progress. After the founding of the People's Republic of China, especially since the reform and opening-up, China realized rapid economic and social development and booming urban and rural construction, resulting in profound changes in the territorial space. The 18th CPC National Congress proposed to "vigorously promote ecological progress" and included "improving the development layout of territorial space" as an important task to promote ecological progress. The *Outline of the 14*th *Five-Year Plan (2021–2025) for National Economic and Social Development and the Long-Range Objectives Through the Year 2035* further specified "optimizing the development and protection layout of territorial space", reinforcing the requirements of respecting, protecting, and reasonable development and utilization of nature.

1 Territorial Space Layout Optimization, An Inevitable Result of Historical Progress

Territorial spatial layout refers to the distribution pattern in a certain geographic space formed by the people in a country or region during territorial spatial development, utilization, protection and improvement activities. Territorial spatial layout is a spatial phenomenon, a historical development process, the mapping of various

J. Shan (✉) · G. Sheng · W. Lou
Research Institute for Eco-civilization, Chinese Academy of Social Sciences (CASS), Beijing, China

Q. Li
University of Science and Technology Beijing, Beijing, China

K. Yang (ed.), *The Construction of Ecological Civilization in China*, China Insights,
https://doi.org/10.1007/978-981-99-6515-1_5

human economic and social activities in geographic space, and the spatial embodiment of national governance strategies. After a long period of effort, Chinese people have completely eradicated absolute poverty and achieved the first centennial goal of completing the building of a moderately prosperous society in all respects in 2020,[1] and China has embarked on a new journey to build a modern socialist country in all respects and achieve the second centennial goal, marking China entering a new era of development and making important adjustments in the development goals and tasks, requiring corresponding changes and optimization of the territorial spatial development layout.

1.1 Development of Territorial Space in the PRC

In more than 70 years since the founding of the People's Republic of China, the territorial spatial development layout and the support system have experienced dramatic changes, in accordance with altering key tasks for national economic and social development at different development stages and major issues facing territorial spatial development. Overall, since 1949, the territorial spatial development in China has roughly gone through four stages.

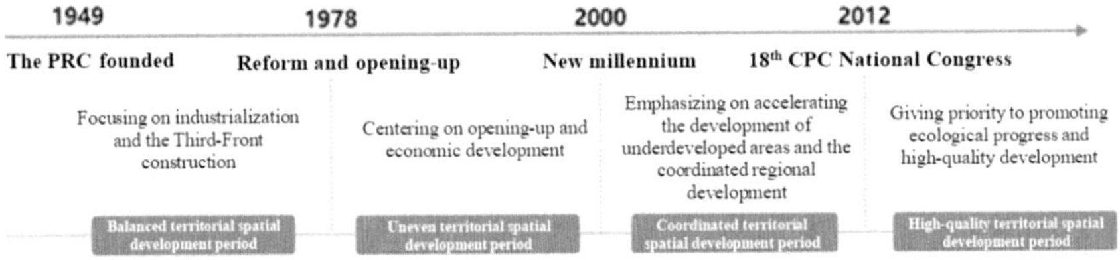

Stages of territorial spatial development in China

In the first period between 1949 and 1977, China maintained a balanced territorial spatial development layout, as it focused on industrialization and the construction of the "Third Front".[2] When just founded, the People's Republic of China had an

---

[1] The report of the 15th CPC National Congress first proposed the two Centennial Goals, namely, completing the building of a moderately prosperous society in all respects by the time the Communist Party of China celebrates its centenary and turning China into a modern socialist country that is prosperous, strong, democratic, culturally advanced, and harmonious by the time the People's Republic of China marks its centennial. Since then, the Two Centennial Goals have become the targets of the Chinese government and the people of all nationalities.

[2] The "Third Front" construction refers to a large-scale industrial relocation process in China against the backdrop of the Sino-Soviet split, U.S. threats to China's southeast coast, and international geopolitical instability. The first "front" was the defense front along the coasts and the borders, the third "front" was the strategic rear of the central and western inland areas, including 13 inland provinces, and the second "front" was the middle area between the first and the third "fronts". For the considerations of national defense and development security, since 1964, China has made large-scale investments in the central and western inland areas, or the "Third Front".

extremely low industrialization level and could not even produce simple industrial products such as electric light bulbs, and the principal contradiction in Chinese society at that time was "the contradiction between the people's demand for an advanced industrial country and the reality of a backwards agricultural country", so accelerating industrialization and altering the development state of enduring impoverishment and long-standing debility became the primary task of territorial spatial development during this period. To this end, the Chinese government required active regional planning and urban planning and a reasonable layout of industrial enterprises and residential areas to establish new industrial zones and new industrial cities and rationally deploy and develop various sectors of industry, energy, transportation, post and telecommunications, agriculture, forestry, and water conservancy, as well as residential areas, while emphasizing that urban construction should serve the industry, production, and residents. During this period, China mainly developed a planned economy by making overall arrangements for limited resources and focusing on industrial development. During the first Five-Year Plan period (1953–1957) alone, China arranged and carried out 694 major construction projects and 156 key projects.[3] During this period, territorial spatial planning mainly manifested in the materialization of national economic plans, and the development of territorial space was mainly the spatial implementation of national economic plans. In the 1960s, due to rapidly deteriorating international environment and international geopolitical environment, national defense security and development security became the main considerations for China's territorial spatial layout. Starting in 1964, China promoted the construction of the "Third Front" and relocated some industries from the coastal and border areas to the central and western inland regions, which sped up industrialization in these inland areas. Therefore, as a whole, China adopted a "balanced strategy" for territorial spatial development, with relatively balanced basic investment in coastal and inland areas, which to a certain extent changed the extremely unbalanced regional industrial development but also brought about problems during that period of inefficient development and increased pressure on resources and the environment in some central and western regions.

In the second period between 1978 and 1999, China strived to promote opening-up and centered on economic development, resulting in unbalanced territorial spatial development. In 1978, China decided to carry out the reform and opening-up and took economic development as the central task. Between 1980 and 1984, the Chinese government established four special economic zones in Shenzhen, Xiamen and other regions and opened 14 cities to the world, including Dalian, Shanghai and Qingdao, aiming at guiding the reform and opening-up in China through piloting in the eastern coastal regions. In addition, the Chinese government also encouraged some regions and people to get prosperous first to achieve common prosperity by boosting others

---

[3] After the founding of the PRC, the Chinese government formulated the first Five-Year plan (1953–1957), arranging 694 projects above the limit, and 921 projects were constructed, of which 156 were built with the aid of the former Soviet Union, known as "156 key projects". Please refer to *Seventy Years of the Communist Party of China* by the Party History Research Center of the CPC Central Committee with Hu Sheng as the chief editor, Chinese Communist Party History Publishing House, 1991, p. 306.

to catch up with the aid from those who already are prosperous. Under the guidance of this development philosophy, China implemented an unbalanced development strategy focusing on the eastern coastal regions, and the allocation of productive forces and resource factors concentrated in the eastern coastal regions promoted the rapid development of these regions, which led to development in China at that time. At the same time, China vigorously advanced market-oriented reforms, implemented the housing commercialization policy in cities, and included the provision that "the right to the use of land may be transferred according to law" in the *Amendment to the Constitution of the People's Republic of China*, while urging decentralization reform between the central and local governments and encouraging delegating administrative and financial powers to local governments, forming a unique decentralization system in China. The reform and opening-up, market-oriented reforms and unbalanced territorial spatial development strategy contributed to China's rapid economic growth but also led to a widening gap in regional economic development and an imbalanced regional development pattern. While large-scale territorial spatial development created economic miracles, it also caused various ecological and environmental problems, pushing the Chinese government to pay attention to spatial governance.

In the third period between 2000 and 2012, China coordinated regional development by focusing on underdeveloped regions. After entering the twenty-first century, to accelerate the development of underdeveloped regions and improve the increasingly serious problem of regional development imbalance, China launched a series of regional development policies, including "Western Development", "Revitalization of Old Industrial Bases in Northeast China" and the "Rise of Central China" in 2000, 2003 and 2004, respectively, turning territorial spatial development from unbalanced to coordinated. During this period, China adopted export-oriented and diversified local economic development models and managed to narrow the development gap between the East and the Central and West for ten consecutive years. However, while regional governments paid more attention "on development than ecology", "on cities than countryside" and "on production than living" and made remarkable achievements in economic development and urban and rural construction, there were problems of low-level repeated construction, damage of different degrees to resources and the environment, and so on, resulting in an unreasonable development layout of territorial space, inefficient use of resources, increased environmental pollution, and degradation of ecological functions in some areas, which directly affected national and regional sustainable development.

In the fourth period from 2012 to the present, China is optimizing the territorial spatial layout, as it targets promoting ecological progress and high-quality development. In 2012, at the 18th CPC National Congress, promoting ecological progress was elevated to a national strategy, and "improving territorial spatial layout" became the primary task for promoting ecological progress. In this period, China will gradually transform the economic and social development model to pursue high-quality development rather than high-speed growth and plan the development and protection of territorial spaces in the new era in a holistic way by coordinating overall considerations concerning population distribution, economic layout, land use and eco-environmental protection and scientifically laying out production space, living space and ecological space. Meanwhile, China will focus on solving problems such as the reduction of high-quality arable land and ecological space and eco-environmental damage caused by disorderly and excessive development, accelerate the formation of a green way of production and life, strive to build a beautiful China and realize the sustainable development of the Chinese nation.

1.2 Demands of the Times for Territorial Spatial Layout Optimization

The world today is experiencing a scale of change unseen in a century, and China has embarked on a new journey to build a modern socialist country in all respects. The major changes in the international and domestic development environment and in the development situation are profoundly affecting territorial spatial development, utilization and governance, not only putting forwards new requirements and posing new challenges to territorial spatial development in China but also bringing new opportunities and directions for the optimization of territorial spatial layout.

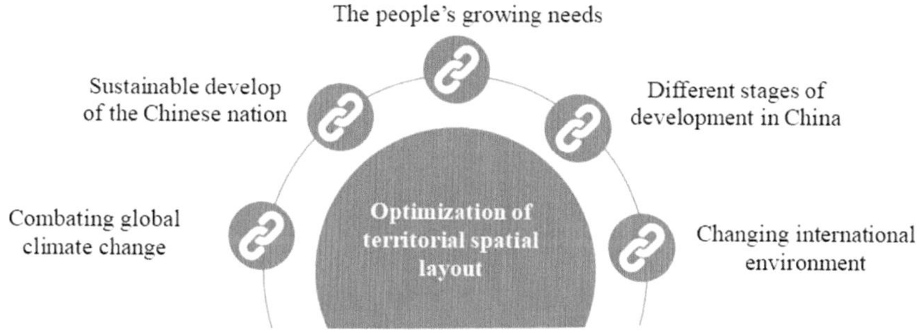

Demands of the times for China to optimize territorial spatial layout

First, changes in the international political and economic environment bring an external impetus for China to optimize its territorial spatial layout. Since the reform and opening-up in 1978, China has long pursued an export-oriented development strategy, and therefore, China generally plans territorial spatial development to support great external economic circulation. However, as the international situation changes, due to factors such as the epidemic of COVID-19, intensified international competition, rising international trade protectionism, and adjustments in global supply chains, international geopolitics become more complex and volatile, and the international market and international development environment become more uncertain and unstable. In addition, the adjustments in global industrial layout accelerate the reshaping of world economic geography, and the dividends of economic globalization are unevenly distributed among countries, regions and even social classes.

All of these factors put forwards higher requirements for territorial spatial development, pushing China to make corresponding adjustments and optimize its original export-oriented development strategy and the territorial spatial development layout.

Second, China enters a new era of socioeconomic development, which brings endogenous impetus to optimize the territorial spatial layout. In 2020, China achieved the first centennial goal of competing to build a moderately prosperous society in all aspects and embarked on a new journey to build a modern socialist country in all respects. In 2021, the urbanization rate in China rose to 64.7%, and China is now slowing the urbanization process rather than accelerating it. The birth rate nationwide is declining as a whole, and the population is expected to reach a peak in 2030 or so. A new round of technological revolution is under way, which is driven by advancements in technologies such as the Internet, big data and artificial intelligence and is having a disruptive impact on people's production methods and lifestyles. And there are many other factors, too. In other words, progress in society, production, economy, science and technology and other fields has brought about effects including "space–time compression", setting new goals for the optimization of territorial spatial layout. The profound changes in the domestic development environment and development situation require China to adjust and improve the territorial spatial development layout correspondingly.

Third, the improvement in people's living needs indicates a new direction for territorial spatial layout optimization. The Chinese government adopts a "people-centered" governance philosophy, emphasizing that "the aspirations of the people to live a better life must always be the focus of our efforts". In 2021, per capita GDP in China exceeded USD 12,500, topping the global average. With economic and social development, Chinese people now have stronger demands for a beautiful ecological environment and high-quality ecological products, and the principal contradiction in Chinese society has evolved and become the contradiction between the people's ever-growing needs for a better life and unbalanced and inadequate development. The people now have a higher living standard, meaning a new direction for "domestic demand" development and new requirements for the optimization of territorial spatial layout. Territorial spatial development must target improving regional quality in a comprehensive and systematic way, creating a better environment for production and living, producing more material and spiritual wealth to meet the growing needs of the people for a better life, and providing more high-quality ecological products on the basis of environmental protection. These requirements will gradually define the new direction for the optimization of territorial spatial layout.

Fourth, China will target the optimization of the territorial spatial layout to secure sustainable development. While China has made great achievements in economic and social development, it has also paid a heavy price in terms of environmental pollution and ecological damage. Meanwhile, global climate change and frequent natural disasters due to the geologically active Earth, have an increasingly strong impact on agricultural production, urban construction, residents' lives and the ecological environment. In addition, facing the current global epidemic, there are more public security emergencies and other security threats, all of which will have an important impact on the development and utilization of national land space. All these factors require China to correspondingly adjust and optimize the development layout, utilization and governance of territorial spaces to guarantee national security and the sustainable development of the Chinese nation.

Fifth, it is a new topic for the optimization of territorial spatial development layout as China is taking part in global climate governance. Global climate governance is an issue with the most far-reaching impact on the world's development and has received the most attention globally. For quite a long time, China has actively participated in global climate governance and advocated for cooperation among countries and regions to combat climate change and protect the Earth, the common homeland. In September 2020, Chinese President Xi Jinping announced at the general debates of the 75th United Nations General Assembly that China would scale up the intended nationally determined contributions (NDCs) by striving to achieve a carbon dioxide emissions peak before 2030 and carbon neutrality before 2060. To fulfil this solemn commitment, China is making efforts to transform the extensive territorial spatial development model, forming a new pattern for territorial development that is green, low-carbon and efficient and enables harmonious coexistence between humans and nature, cooperate with international communities to combat great challenges, including climate change, and contribute to the construction of a beautiful, prosperous and sustainable new world.

2 Territorial Space Layout Optimization, Targeting at Fulfilling the Chinese Dream

The Chinese government stresses that the construction of ecological civilization is a national project of millennial significance to sustainable development of the Chinese nation and an important part of the socialist cause with Chinese characteristics, while the territorial space is the spatial carrier for the ecological civilization, so optimizing the territorial spatial layout is the primary task to promote ecological progress. By promoting the coordination of production, living and ecological spaces and building a safe, harmonious and competitive territorial spatial layout that is conducive to sustainable development and ensures that space for production is used intensively and efficiently, that the living space is liveable and proper in size and that the ecological space is unspoiled and beautiful, it means providing the material basis and spatially sustaining the construction of a beautiful China and the cause of socialist modernization and making new contributions to global ecological progress and security.

In the new era, China will drive the optimization of the territorial spatial layout to achieve the following goals.

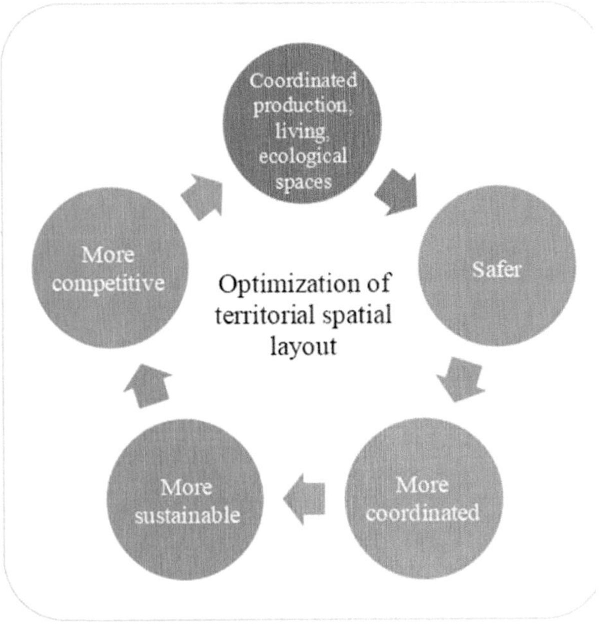

Goals of China's optimization of territorial spatial layout

2.1 *Coordinated Production, Living and Ecological Territorial Spaces*

The Chinese government has always emphasized the need to promote the coordinated development of spaces for production, living and ecology and to advance urbanization, agricultural development and ecological security in a scientific and balanced way. For urbanization, China insists on intensive land development and effectively improves the intensive use of urban construction land by strictly controlling and optimizing additional land development and perfecting the use of land available and the land development structure for higher efficiency and quality utilization to construct safe and liveable cities. For agricultural production, China focuses on effectively protecting agricultural spaces such as arable land, garden land and vegetable land. For eco-environmental protection and construction, China will delineate and strictly observe the ecological red lines to "leave to our future generations a beautiful homeland with blue sky, green fields and clean water".[4]

[4] Xi Jinping, *Xi Jinping: The Governance of China*, Vol. I, Foreign Languages Press, 2014, p.212.

2.2 Constructing a Safe Territorial Space

With the development of the economy and society, China is now facing an increasing number of security issues, including security for ecology, energy, food, finance, networks, information and so on. The Chinese government emphasizes that these security issues in development can only be resolved through comprehensive economic, social, and technological improvements. It needs to "attach importance to both development and security issues, since development is the basis for security and security is the premise for development", to integrate development and security, while "emphasizing both its own and common security" to promote international security and discard the zero-sum thinking, "promoting all parties to move toward the goal of mutual benefit and common security" and build a global community of shared future. The construction of a safe territorial space is an important basis for maintaining national and ecological security as well as an important task for optimizing the territorial spatial layout.

2.3 Building a Harmonious Territorial Space

To realize the goal, first, China should uphold the harmonious coexistence between humans and nature, follow the principle that human economic and social development activities must respect, accommodate and protect nature, and place ecological conservation and environmental protection in a more prominent position to "protect the ecological environment like maintaining our health". Second, China should promote coordinated regional development and achieve equitable access to basic public services to ensure that people in different regions have roughly equal living standards and eventually achieve common prosperity. Finally, China should facilitate the integrated development of urban and rural areas, advance urbanization and rural revitalization in a holistic way, and develop a new type of relationship between industry and agriculture and between urban and rural areas in which industry promotes agriculture, urban areas support rural development, agriculture and industry benefit each other, and urban and rural areas develop together.

2.4 Developing a Competitive Territorial Space

At present, China is making every effort to promote socialist, so it must put quality first, give priority to performance, continue to improve quality and efficiency, strengthen the driving forces of economic development, raise total factor productivity, and steadily enhance economic innovation capacity and competitiveness to shape a more competitive territorial space.

2.5 Improving the Sustainability of the Territorial Space

China holds the view that "lucid water and lush mountains are invaluable assets, that protecting the environment means protecting productive forces, and that improving the environment equals developing productive forces" and requires that "resource development and utilization should not only support current generation to lead a happy life but also lay solid a foundation for future generations to survive." To this end, China must develop the territorial space, industrial structures, and modes of production and living that are conducive to the conservation of resources and the protection of the environment, give time and space for natural ecosystems to rehabilitate themselves, and seek the road to sustainable development by upholding the traditional Chinese philosophy of "Man is an integral part of nature, and Dao follows the laws of nature".

3 Laying a Foundation for Sustainable Development Through Practices and Innovations in Territorial Space Layout Optimization

Based on the construction of ecological civilization, a beautiful China and socialist modernization, China aims to develop a safe and harmonious, competitive and sustainable territorial spatial layout with coordinated spaces for production, living and ecology, innovate the territorial spatial governance system for ecological progress, advance the implementation of a territorial spatial optimization strategy, promote river basin-based ecological progress, and focus on building a territorial spatial system that sustains high-quality development.

3.1 Innovating the Territorial Spatial Governance System

Since the 18th CPC National Congress, ecological civilization has become a national development strategy and the center of reform for China's territorial spatial governance system. In 2012, the report of the 18th CPC National Congress set "building a beautiful China" as the target for promoting ecological progress. In 2015, China issued the *Integrated Reform Plan for Promoting Ecological Progress*, clarifying that China will boost the reform to promote ecological progress mainly by improving the functional zoning system and establishing a planning system for territorial space. In 2018, the Ministry of Natural Resources of China was established to solve problems such as "inexplicit ownership of natural resources and overlapping spatial planning" and to "perform in unified way the duties of regulating the use of all territorial space and protecting and restoring ecosystems", gradually forming a territorial spatial governance system centering on ecological progress, focusing on functional zoning and territorial spatial planning, and targeting at "building a beautiful China".

First, China needs a planning system for functional zones. In 2010, the State Council issued the *National Plan for Developing Priority Zones* to form a new development layout with explicitly defined regional functions that is conducive to complementing regional advantages and achieving overall high-quality development. The functional zone planning means to coordinate overall considerations concerning population distribution, economic layout, land use and urbanization modes based on the resource and environment capacity, existing development intensity and development potential of different regions, determine the primary functions of different regions and the direction for development, and then improve the policy, control the intensity and regulate the order for development to gradually form a territorial spatial development layout accommodating to local population, economy and resource environment. Functional zones in China are divided into two levels and two kinds, namely, national and provincial zones and terrestrial and marine zones. In addition, according to whether it is allowable for large-scale and high-intensity industrialization and urbanization development or the development modes it allows, the territorial space is divided into areas of four types: optimized development areas, key development areas, restricted development areas and prohibited development areas. According to the content of development and the main functions (the products provided by the region), the terrestrial territorial space is divided into urban areas, primary production areas for agricultural products, and key ecosystem service areas, while the marine territorial space is divided into industrial and urban construction areas, agricultural and fishery production areas, and ecological and environmental service areas.

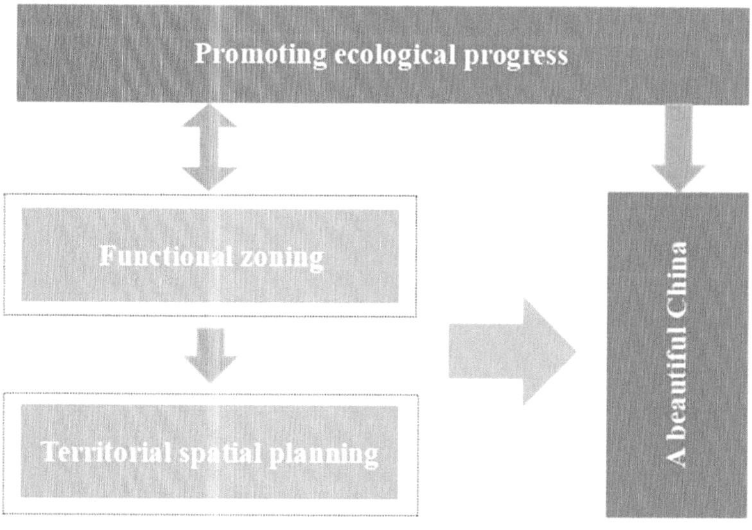

Territorial Spatial governance system for ecological civilization

Second, China needs to innovate the territorial spatial planning system. Territorial spatial planning provides guides for territorial development, and it is the spatial blueprint for sustainable development and the guidelines for various development, protection and construction activities. In 2019, China issued *Several Opinions of the CPC Central Committee and the State Council on Establishing a Territorial Spatial Planning System and Supervision on Implementation*, specifying that "territorial spatial planning is the spatial and temporal arrangements for the development and protection of territorial space in a certain area, including master plans, detailed plans and relevant special plans". In China, there is an administrative hierarchy system of five levels, and it is required that the master territorial spatial plans have to be prepared at national, provincial, municipal and county levels and that township territorial spatial plans should be developed based on actual regional conditions, marking that the system of territorial spatial planning composed of plans at "five levels" and of "three types" has been formally established.

National territorial spatial planning is the overall arrangement for national territorial space, and it is strategic, as it provides a policy and general principle for the protection, development, utilization and restoration of national territorial spaces. Provincial territorial spatial planning is the implementation of national territorial spatial plans, and it is coordinating since it guides the preparation of municipal and county territorial spatial plans.

Five-level and three-type territorial spatial planning system in China

Territorial spatial planning at cities, counties and townships means the detailed implementation of the requirements of superior territorial spatial planning by governments at these levels and the specific arrangements for territorial spatial development and protection in these administrative regions, so such plans are practical.

Detailed plans are prepared at cities, counties and lower administrative levels, and they are the implementation arrangements for a specific plot of land, including the purpose, development and construction intensity, etc., providing a legal basis for implementing territorial spatial development and protection activities, regulating the use of the territorial space, issuing permits for urban and rural construction projects, carrying out various constructions, and so on.

There are two types of special plans. The first type includes territorial spatial plans for coastal belts, nature reserves, basins, and other cross-administrative or nonadministrative regions, while the second consists of special plans for certain utilization of the territorial space, such as development plans in the transportation, energy, water conservancy, and ecological environmental protection sectors. Special plans are important constraints on territorial spatial planning.

3.2 *Improving the Development of Functional Zones*

In 2010, China issued the *National Plan for Developing Priority Zones*, which divides the territorial space of China into areas of four types according to the development modes, namely, optimized development areas, key development areas, restricted development areas and prohibited development areas, and areas of three types according to the development purposes, that is, urban areas, primary production areas for agricultural products and key ecological service areas. Currently, there are 25 national key ecological service areas in China, covering 436 county-level and 22 provincial administrative regions, and 1,443 prohibited development areas have been delineated, with a total area of 1.2 million square kilometers, accounting for 12.5% of the land mass (Table 1).

Meanwhile, based on the development intensity and the distribution of development areas, China also develops strategic territorial spatial development layouts: an urban development pattern of "two horizontal and three longitudinal lines", an agricultural production pattern of "seven areas and 23 belts" and an ecological security pattern of "three areas and four belts" (Table 2).

In addition, China actively promotes the construction of marine functional zones. In August 2015, China issued the *National Plan for Developing Marine Priority Zones*, which divides marine function zones into three types according to their development purposes: areas for industry and town construction, areas for agricultural and fishery production, and ecological and environmental service areas. In addition, China divides the marine territorial space into areas of four types based on the functions: optimized development areas, key development areas, restricted development areas and prohibited development areas. As of 2018, all coastal provinces in the Chinese mainland had completed the preparation of marine functional zone plans, delineating a total of approximately 104,000 square kilometers of marine optimized development areas, approximately 15,500 square kilometers of key development areas, approximately 126,000 square kilometers of restricted development areas, and approximately 25,000 square kilometers of prohibited development areas,[5] which marked that the national functional zoning strategy in China had full coverage of the terrestrial and marine territorial space.

[5] Data based on these provincial plans for marine functional zones.

Table 1 National key ecological service areas and prohibited development areas in China

Type	Specific areas
(25) national key ecological function zones	Daxing'anling and Xiaoxing'anling Forest Ecological Function Zone, Changbai Mountain Forest Ecological Function Zone, Mountain Forest and Grassland Ecological Function Zone in Altai Mountain, Sanjiangyuan Grassland, Marshy Grassland and Wetland Ecological Function Zone, Ruoergai Grassland and Wetland Ecological Function Zone, the Ecological Function Zone in Southern Gansu Province of an Important Water Replenishment Region for the Yellow River, Qilian Mountains Glacier and Water Conservation Ecological Function Zone, Nanling Mountain Forest and Biodiversity Ecological Function Zone, Hill-gully Area Soil and Water Conservation Ecological Function Zone on the Loess Plateau, Dabieshan Mountain Soil and Water Conservation Ecological Function Zone, Guangxi- Guizhou-Yunnan Karst Desertification Control Ecological Function Zone, Three Gorges Reservoir Area Soil and Water Conservation Ecological Function Zone, Tarim River Desertification Control Ecological Function Zone, Aerjin Grassland Desertification Control Ecological Function Zone, Hulun Buir Grassland and Marshy Grassland Ecological Function Zone, Keerqin Grassland Ecological Function Zone, Hunshandake Desertification Control Ecological Function Zone, the Grassland Ecological Function Zone in the North Root of Yinshan Mountain, Sichuan and Yunnan Forest and Biodiversity Ecological Function Zone, Qinba Biodiversity Ecological Function Zone, the Forest Ecological Function Zone on the Edge of Qinghai-Xizang Plateau in southeastern Xizang, Qiangtang Plateau Desert Ecological Function Zone in northwest Xizang, Sanjiangyuan Plain Wetland Ecological Function Zone, Wuling Mountain Biodiversity and Soil and Water Conservation Ecological Function Zone, the Tropical Rainforest Ecological Function Zone in Central Mountainous Areas of Hainan Island
(1,443) national prohibited development areas	(319) national nature reserves, (40) world cultural and nature heritages, (208) national scenic areas, (738) national forest parks, (138) national geological parks

To implement and effectively promote functional zone planning, China has designed a "9 + 1" policy system, of which "9" refers to policies in the nine sectors of finance, investment, industry, land, agriculture, population, nationality, environment, and climate change, and "1" means the performance evaluation and assessment policy. For example, to compensate these regions for the loss of economic development Opportunities because the planned functional zones are required to provide ecological products and protect the ecological environment, the central and local governments have gradually established an ecology-related financial transfer payment system. All 451 counties affected, meaning the ecologically restricted and prohibited development areas delineated by the *National Plan for Developing Priority Zones* are in these counties, have been covered by the national financial transfer payment system. From 2017 to 2021, the amount of transfer payments for

Table 2 Territorial spatial development layout in China

Type	Overall layout	Specific areas
Strategic urban development pattern	Two horizontal and three longitudinal lines	**"Two horizontal lines"**: the region along the Land Bridge Passage, the region along the Yangtze River **"Three longitudinal lines":** the coastal region, the region along Haerbin-Beijing-Guangzhou railway, the region along Baotou-Kunming High Speed Railway
Strategic agricultural production pattern	Seven areas and 23 belts	**"Seven areas"**: the Northeast China Plain, the Huanghuaihai Plain, the Yangtze River Basin, the Fenwei Plain, Hetao Irrigation Area, the South China, Gansu Province and Xinjiang Uygur Autonomous Region, seven primary production areas for agricultural products. **"23 belts"**: the industrial belts of high-quality rice, corn for special end use, soybeans and livestock products in the Northeast Plain primary production area for agricultural products; the industrial belts of high-quality wheat for special end use, high-quality cotton, corn for special end use, soybeans and livestock products in the Huanghuaihai Plain primary production area for agricultural products; the industrial belts of high-quality rice, high-quality wheat for special end use, high-quality cotton, oilseed rape, livestock and aquatic products in the Yangtze River Basin primary production area for agricultural products; the industrial belts of high-quality wheat for special end use and corn for special end use in Fenwei Plain primary production area for agricultural products; the industrial belts of high-quality wheat for special end use in Hetao Irrigation Area primary production area for agricultural products; the industrial belts of high-quality rice, sugar cane and aquatic products in South China primary production area for agricultural products; the industrial belts of high-quality wheat for special end use and high-quality cotton in Gansu and Xinjiang primary production areas for agricultural products
Strategic ecological security pattern	Three areas and four belts	**"Three areas"**: Qinghai-Xizang Plateau ecological shield zone, the Yellow River key ecological zone, the Yangtze River key ecological zone **"Four belts"**: the northeast China forest belt, the northern China desertification-prevention belt, the southern China mountainous belt, and the coastal belt

national key ecological functional zones increased from USD 8.68 billion to USD 12.05 billion. In addition, the central government has established special ecological compensation mechanisms for seven ecological areas, including grasslands, wetlands, and forests, and between 2010 and 2020, the central government cumulatively paid USD 249.12 billion of various ecological compensation.[6] Meanwhile, more than 20 provincial governments have introduced ecological compensation policies for river basins, and 10 provincial governments have introduced ecological compensation mechanisms at lower administrative levels for key function zones.

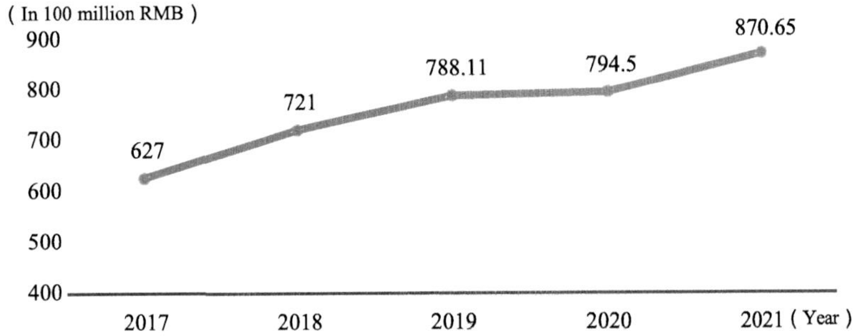

Changes in the transfer payments for national key ecological function zones

Data source the Ministry of Finance of the People's Republic of China, http://www.mof.gov.cn/index.htm

3.3 *Implementing Strategies for Territorial Spatial Layout Optimization*

First, China should foster a new development pattern featuring mutually reinforcing domestic and international dual circulation. For a long time, China was a "world factory" at the lower end of the global industrial chain, with a low added value and a weak brand effect. In addition, once-in-a-century world change, profound adjustments in the global economic landscape, the continuously weak international market, and the trend of partial "decoupling" of the industrial and supply chain from China all pose new challenges to economic development in China. In the face of the still raging epidemic of COVID-19 and complex international economic situation, China, from the perspective of building a community with a shared future for mankind, proposed to keep in mind both the internal and international imperatives and give full play to the advantages of the mega-market in China and the huge potential in domestic demand to build a new dual circulation development pattern in which domestic and

[6] Data Source: The Ministry of Finance of the People's Republic of China, http://www.mof.gov.cn/index.htm.

international circulation can complement and reinforce each other, to cultivate new drives of growth and open up new prospects by building the new development pattern and to promote the continuous optimization of the territorial spatial layout.

Second, China needs to foster a regional economic layout that enables complementary and high-quality development. At present, China is exploring a new economic development model focusing on quality rather than growth rate, as in China, there are new situations in regional development, such as the prominent differentiation between North and South China and the polarization phenomenon of driving forces. In the face of these new situations and changes, the Chinese government proposed a new idea of adapting to the new situation and planning for coordinated regional development: by taking a basis on the carrying capacity of resources and the environment and giving full play to the comparative advantages of each region, China should promote the reasonable flow and efficient concentration of various factors, form different policy units according to the functions of various functional zones, and formulate and carry out differentiated policies for key development areas, areas rich in energy resources, ecological shield zones, and so on, to foster a new territorial spatial development and protection layout with explicitly defined functions that is conducive to complementary and high-quality development.

Third, China needs to actively develop new momentum for high-quality development. For example, China is vigorously advancing the construction of the Beijing-Tianjin-Hebei Urban Agglomeration and the city cluster in the Yangtze River Delta and developing the Guangdong-Hong Kong-Macao Greater Bay Area to improve the capacity in global resource allocation and in creational and original innovation and to set up examples that could guide China's high-quality development. In addition, guided by key cities, China is boosting the development of central and western China, including the twin-city metropolitan area of Chengdu and Chongqing and the city cluster in the middle reaches of the Yangtze River, to accelerate industrialization and urbanization in regions where conditions are available and cultivate high-quality regional development. China is fully cultivating and giving full play to the role of city clusters and metropolitan areas in stimulating and supporting the surrounding regions to gradually enhance the comprehensive carrying capacity, form a development network starting from a single point, and drive the overall improvement in economic efficiency in China.

In 2019, the CPC Central Committee and the State Council issued the *Outline for Regional Integrated Development Plan of the Yangtze River Delta* and decided to develop the Yangtze River Delta into a high-level demonstration zone for green and integrated ecological development. The construction of the Demonstration Zone of Green and Integrated Ecological Development of the Yangtze River Delta breaks the geographic boundaries of the traditional governance model for territorial resources based on administrative regions and helps develop a new governance pattern by unifying and coordinating regional policies, planning, standards and norms based on the natural links formed by the Yangtze River and the river system, solving problems such as "concerning one's own business" or "being nobody's business", and achieving common development and management of territorial resources through negotiation, as well as sharing of achievements. In the three years from 2018 to 2021, the ecological governance of Yuandang Lake made remarkable achievements. The indexes of water quality in Yuandang Lake were improved significantly, with the amount of ammonia nitrogen and total nitrogen being reduced by 49% and 36%, respectively, meaning that the water in Yuandang Lake meets the Grade IV water quality standard on most days a year, and the lake that used to be neglected has been restored ecologically. Similarly, thanks to the joint prevention and treatment mechanism, the number of days with good quality in the Beijing-Tianjin-Hebei Urban Agglomeration reached 260, and in 2021, the average PM2.5 concentration was 36.9 mcg/m^3, down 62.3% compared to 2013. By actively playing the advantages in science, technology and innovation, the Guangdong, Hong Kong and Macao Greater Bay Area scientifically solved the problem of ecological constraints on economic development and constantly improved the ecological and environmental conditions. According to the report on the work of the Guangdong government issued in 2022, a high-quality clear waterway of 2,075 km was built, 1.92 million *mu* (316, 295 acres) of land was afforested or remediated, and 693 hectares of mining areas turned green in Guangdong in 2021, while 90.2% of coastal and sea water were rated as excellent or moderate quality.

3.4 Improving the Regulatory System for the Use of Territorial Space

First, China needs to improve the regulatory system for territorial spatial development. Early in the 1980s, China began to pay attention to the regulation of territorial space, but there were problems of too many regulating bodies, overlapping regulatory functions and uncoordinated regulations since the territorial spatial planning functions were scattered among different departments, including development and reform, land and resource management, urban and rural construction, and environmental protection. To this end, in 2013, the Chinese government proposed to "demarcate the development boundaries of production, living and ecological space, carry out the regulatory system for the use of territorial space", and "improve the system for regulating natural resources and perform all use-related regulatory duties and responsibilities for all territorial spaces".

In 2017, China issued the *Measures for Managing Natural and Ecological Space* (For Trial Implementation) and selected six provinces of Fujian, Jiangxi, Guizhou, Henan, Hainan and Qinghai to pilot the regulation on the use of natural and ecological space at the provincial, municipal and county levels. In addition, subscribing to the philosophy that "mountains, rivers, forests, fields, lakes and grasslands are part of a community of life", China coordinated various regulatory systems for spatial use and gradually extended the regulatory scope to cover all natural and ecological spaces. In 2018, the Chinese government established the Ministry of Natural Resources to perform all use-related regulatory duties and responsibilities for all territorial spaces and to actively promote the detailed implementation of the national regulatory system for the use of territorial space by combining the "top-down" and "bottom-up" approaches. Located in the Qilian Mountains, Menyuan Hui Autonomous County, Haibei Tibetan Autonomous Prefecture, Qinghai Province, faced relatively severe local eco-environmental problems. In May 2017, the state approved including the Qilian Mountains in Qinghai Province in the pilot integrated ecological protection and restoration project for mountains, rivers, forests, fields, lakes and grasslands, replacing the previous projects special for one purpose, such as mountain protection, water conservation and arable land protection projects. Menyuan County adheres to the laws of nature, carries out territorial spatial protection in accordance with local conditions, namely, planting trees or grasses on mountains based on climate conditions, and develops various ecological conservation and compensation mechanisms to achieve innovative ecological value transformation and the engagement of the whole society. During the Fourth Session of the 13th National People's Congress in 2021, two photos were shown to President Xi Jinping, which best displayed the achievements of the integrated governance of mountains, rivers, forests, fields, lakes, and grasslands. The photos of snow leopard and Chinese Mountain Cat were taken in Qilian Mountains National Park.

Second, China needs to coordinate the delineation and implementation of three control lines. To promote ecological progress and improve the regulation of territorial spatial development, early in the 1990s, China began to draw red lines for ecological conservation and delineate permanent basic cropland and boundaries for urban development. In 2019, the delineation of the "three lines" was unified into the territorial spatial planning system. The permanent basic cropland was delineated earliest. It was calculated that China needs 1.8 billion *mu* (296.53 million acres) of arable land to ensure food security. To ensure that the area of arable land stayed above the "red line" of 1.8 billion *mu*, the Chinese government proposed for the first time in 2008 the "delineation of permanent basic cropland". In 2011, China launched the delineation and adjustment of basic cropland in cities, counties and townships, and by 2017, all 2,887 county-level administrative districts had finished the delineation of specific plots, marking the full completion of delineating permanent basic cropland.

Following the declaration of a red line of no less than 1.8 billion *mu* of arable land, the Chinese government drew red lines for ecological conservation as another national lifeline. In 2012, the Ministry of Environmental Protection (MEP) of China prepared a preliminary draft of the *Technical Guide for Delineating National Red Lines for Ecological Conservation*, and in 2013, China began to pilot the delineation of redlines in the Inner Mongolia Autonomous Region, Jiangxi Province, Hubei Province, Guangxi Zhuang Autonomous Region and other provincial administrative regions. By the end of 2020, the delineation of national red lines for ecological conservation was basically completed, and governments at all levels began to shift attention to the supervision of these ecological red lines.

The study of boundaries for urban development started relatively late in China. In 2014, the Chinese government identified 14 pilot cities, including Beijing, Shenyang, Shanghai and Nanjing, to pilot delineate urban development boundaries, and in 2015, more than 600 cities nationwide began delineating urban development boundaries. In 2022, as the cities across China formulated master plans for territorial space, cities and towns delineated the boundaries for urban development one after another. By drawing and strictly adhering to the red lines for ecological conservation and permanent basic cropland and boundaries for urban development, China aims to optimize the spaces for production, living and ecology and guide the formation of a scientific, moderate and orderly territorial spatial layout.

Biyang County in Henan Province is an ordinary county in China and one of the third batches of "demonstration counties for building ecological civilization in China". Centering on coordinated spaces for production, living and ecology, Biyang County insists on advancing the development, regulation and governance of production, living and ecological spaces harmoniously while promoting the work of river cleaning, rural sewage treatment, livestock breeding and so on in a holistic way and has made remarkable achievements in both economy and ecology: 100% of water at drinking water sources in the county and 100% of cross-sections of surface water flowing out of the county boundaries met relevant standards. Public participation in building ecological civilization in Biyang County reached 90.7% and 93.8% for the degree of satisfaction. From Beijing and Shanghai to other cities and towns in China, projects focusing on the integration of industry and a city or town have constructed many ecological and industrial zones with coordinated spaces for production, living and ecology.

3.5 Strengthening the Ecological Progress Based on River Basins

River basins are usually the birthplaces and centers of human civilizations and are also spatial carriers and basic elements for promoting the harmonious coexistence between humans and nature. China is a vast country covering a land area of 9.6 million square kilometres, and there are 228 rivers with a basin area of more than 10,000 square kilometres and more than 2,200 rivers with a basin area of more than 1,000 square kilometers, and nearly 23,000 rivers with a basin area of more than 100 square kilometers. Diversified basins have the largest population and generate the most economic output, while they are the most important treasury of biodiversity and shields for ecological security.

As industrialization advanced, the water environment in China began to deteriorate regionally in the 1980s and as a whole in the 1990s. In 2000, at the Fifth Plenary Session of the 15th CPC Central Committee, it was emphasized that "sustainable utilization of water resource is a strategic issue concerning economic and social development", and China called for strengthening water pollution control and water ecological restoration. To this end, China has formulated and implemented five successive five-year plans for water pollution prevention and control in major river basins, and in 2017, the scope of these plans was extended to cover all key river basins in China. Chinese President Xi Jinping attaches great importance to the territorial spatial development of river basins and has visited the basins of the Yangtze

River, the Yellow River and other major rivers several times, emphasizing the need for systematic governance and coordinated protection of the "upstream and downstream, trunk and tributaries, left and right banks" of these rivers. Meanwhile, he has personally promoted ecological progress in the Yangtze and Yellow River basins.

Taking the Yangtze River as an example, it is the largest river in China and Asia and the third largest river in the world, with a total length of 6,380 km, and it runs through 19 provincial administrative regions in eastern, central and western China. The Yangtze River basin covers approximately 1/5 of China's land area, sustains approximately 1/3 of the population, outputs 1/3 of the food, and generates 1/3 of the GDP in China. To protect such an important river and the lifeblood of China's economic development, the Chinese government must "adhere to ecological priority and green development", "step up conservation and stop over development", and "put the ecological restoration of the Yangtze River in an overriding position" while carrying out a series of specific actions. First, China is taking efforts to win the battle to improve the water quality. Targeting at industrial, domestic sewage, rural nonpoint source pollution and other prominent problems that damage the ecological environment of the Yangtze River Basin, China launched vigorous special rectification actions, requiring these enterprises with backwards technology or insufficient environmental protection facilities or failing to meet national environmental standards to "shut down, suspend the operation, merge or switch production", while promoting industrial structural optimization and upgrading and the development of clean industry and service industry to create more jobs for local people. Second, China continues to strengthen coordinated ecological and environmental governance across the basin region and has issued a series of river basin-based laws, plans and action plans for water ecological conservation and environmental protection, requiring cooperation between governments of different administrative regions to strengthen the coordinated eco-environmental governance of the entire basin area. Third, China pays more attention to building the biotic community of mountains, rivers, forests, fields, lakes and grasslands to reinforce the protection and restoration of basin ecosystems. Through systematic governance, many regions used to be seriously polluted, such as Wuhan Donghu Lake and Hangzhou Xixi Wetland, and have already rebuilt healthy ecosystems.

After systematic environmental governance and ecological restoration between 2015 and 2020, the proportion of water sections with surface water quality meeting Grade I and II standards increased by 4.4 and 12.8% points, and the proportion of those meeting Grade III, IV and V standards and failing to meet Grade V standards declined by 10, 3.3, 0.8, and 3.1% point(s), respectively. It realized a historical achievement that 100% of the trunk water of the Yangtze River was rated as excellent quality. Meanwhile, the combined proportions of water sections in major river basins in China with surface water quality reaching Grades I and II increased by 5.1 and 13.7% points, respectively, and the proportions of those meeting Grade V and failing to meet Grade V decreased by 3.2 and 8.7% points, respectively, marking the overall quality of the water environment in each river basin improved, and China has dramatically promoted ecological progress in river basins. In other words, China is becoming more beautiful, as the spaces for production, living and ecology are coordinated, the rivers are cleaner, the mountains are greener, and the people are happier.

Donghu Lake of Wuhan (the picture above) and Xixi Wetland (the picture below) in Hangzhou after the integrated governance of mountains, rivers, forests, fields, lakes and grasslands

Source Qiushi, "China's Path to Ecological Civilization", http://www.qstheory.cn/dukan/qs/2019-11/01/c_1125178887.htm

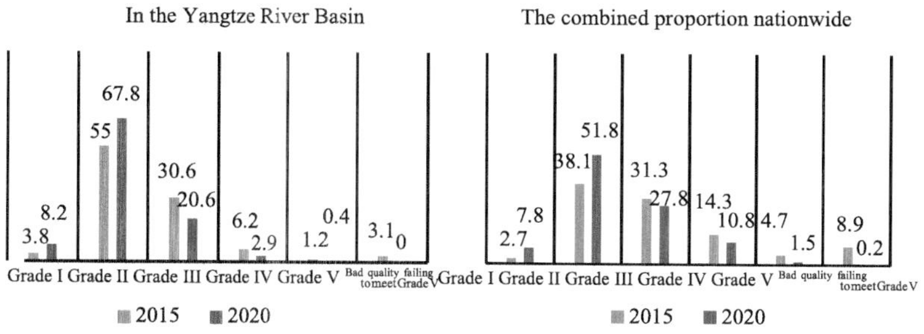

Changes in the proportions of water sections with various surface water qualities in the Yangtze River Basin and other basins in China between 2015 and 2020

Source Data based on Bulletin on the State of China's Ecological Environment for 2015 and 2020

Chapter 6
Jointly Building a Shared Future for Humans and Nature

Xiang Yu

2021 was a critical year, as the whole world expected great achievements would be made and cooperation on combating global climate change would be strengthened, especially between China and the United States, as both are important major powers to address climate change and promote global cooperation on climate issues. On April 22nd, 2021, the 52nd World Earth Day, at the invitation of U.S. President Joe Biden, Chinese President Xi Jinping, attended the Leaders Summit on Climate and delivered a speech via video link from Beijing. In the speech titled *For Man and Nature: Building a Community of Life Together*, President Xi Jinping comprehensively elaborated the concept of governance featuring harmonious coexistence between humans and nature from the perspective of global ecological security beyond racial, cultural, consciousness and national boundaries. He put forwards the development principles of committing to the harmony between humans and nature, green development, systemic governance, people-centeredness, multilateralism, and common but differentiated responsibilities and demonstrated to the world the strategy and path for China to join hands with other countries to foster a new relationship where humans and nature can live in harmony. Balancing the relationship between humans and nature during the development process is the primary issue that must be faced and resolved for all human civilizations. Nature has no boundary, and it will be a global practice beyond the constraints of times, territories, politics and nationalities to promote the building of a community of harmony between humans and nature. China is courageous as a pioneer and pragmatic in the actions to provide the world with Chinese ideas, wisdom and experience in promoting the building of a shared future for humans and nature.

X. Yu (✉)
Research Institute for Eco-Civilization, Chinese Academy of Social Sciences (CASS), Beijing, China

K. Yang (ed.), *The Construction of Ecological Civilization in China*, China Insights,
https://doi.org/10.1007/978-981-99-6515-1_6

1 Rich Implications of Jointly Building a Shared Future for Humans and Nature

1.1 Interdependent Humans and Nature Sharing the Same Future

On April 22nd, 1970, the first "Earth Day" activities were held. However, at present, the climate and environmental problems faced by humans are still serious: continuously rising global temperature, environmental pollution, desertification, losses of soil and water, accelerating extinction of biospecies, and frequent extreme weather events such as acid rain, severe rainfall, typhoons, hurricanes, and tsunamis, all posing great challenges to human survival and development. The global epidemic of COVID-19 broke out in 2020 and has not been effectively controlled thus far, threatening public health and security sectors in countries and regions around the world. As the epidemic continues to rage, the global economy shows a recessionary trend while the clouds of war persist. From a certain point of view, human society is in a hard period of multiple crises, all of which, to some extent, stem from the improper handling or mismanagement of the relationship between humans and nature.

Chinese civilization has a glorious history of several thousand years and developed sustainable agrarian models precisely because of the philosophy of respecting, accommodating and rationally utilizing nature and improving natural productivity, resulting in several powerful and prosperous dynasties. In ancient China, although the understanding of nature was limited, profound ecological wisdom and philosophical thinking emerged as "nature is the true law" and "when humans and nature are in harmony, a beautiful homeland will form", reflecting in a simple way the wisdom of dealing with the relationship between humans and nature. In the pre-Qin period as early as approximately 400 BC, it was stated in the chapter "Lizheng (立 政, Establishing the Governance Principals)" of the significant work of ideology *Guanzi* that "山泽救于 火, 草木植成, 国之富也", meaning that if mountains and forests are effectively protected from disasters, including fires, grasslands and rivers, are conserved, the country and society will develop steadily and prosper. These ideas bear a similar philosophy and clearly illustrate from different perspectives that the use of ecological and biological resources by humans should follow the laws of nature and that the rational use and protection of resources is the basis for human survival and national prosperity. The ecological concepts of accommodating to, respecting and protecting nature, and sustaining resources, which were summarized by visionary Chinese thinkers in the past based on production and living practices, carry a simple development concept of paying attention to the balance between ecological protection and social (national) governance and have been passed down to now as a part of the 5,000-year Chinese civilization, exerting far-reaching effects. Although the ancient Babylonian and Mayan civilizations had the glory of conquering nature, the deteriorated ecological environment due to the violation of natural laws eventually led to the decline and even the extinction of these civilizations. The civilization process of human history tells us that the use of natural resources for economic development must be "moderate" and "in order" to achieve harmony with nature. Otherwise, nature will "revenge" the "conquerors". Civilizations based on the conquering of nature cannot last for long since there is no sustainable resource supply for such prosperity.

1.2 Nature Laying Foundations for Human Society and Sustaining Development and Prosperity

With the development of human society, the economic value of natural assets has become increasingly prominent. Not only has the natural environment become an indispensable element for human well-being, but ecological industries, such as eco-cultural industries, eco-agriculture, eco-industry and eco-tourism, are making environmental contributions to economic development and promoting sustainable social development. Since ancient times, there in China has been a saying that "先王之法, 不涸泽而渔, 不焚林而猎", demonstrating that the ancient Chinese have realized that as long as natural resources are not overexploited, nature, to a certain extent, can restore and heal itself, "preserving" and "increasing" its value and making more economic contributions to human development. In various dynasties in China, there were laws stipulating that hunting and killing cubs or entire animal populations, burning forests or drain ponds for hunting or fishing were not allowed, and there were specific moratorium systems for hunting and fishing, requiring people to respect the laws of natural reproduction and growth of all kinds of wildlife and to choose the most economically efficient time to carry out production activities in a reasonable and proportionate manner to meet the development needs of human society. Natural resources are a natural source of production factors for a country's prosperity, while a beautiful environment is a treasury for people to seek spiritual wealth. Even today, many urban citizens in modern metropolises prefer hiking and excursions for tourism, intuitively reflecting human desires to return to and feel nature. The Chinese educator Confucius' ecological ethical remarks of "understanding the responsibilities and fearing the laws of nature (知命畏天)" and "the benevolent love mountains, while the wise love water (知者乐水, 仁者乐山)", came from the spiritual enjoyment and dematerialized value brought by nature. In addition, birds, animals, insects and fish, and plants and flowers have been important subjects and contents in traditional Chinese art since ancient times. The images of natural life in Chinese art works and the spirit of life they embody manifest the attitudes of traditional Chinese culture toward nature and life, reflecting the all-round harmony between humans and nature and human respect and love for nature. Natural elements are important carriers in a large number of Chinese literary works to express virtues, sentiments and ideals, and nature is also an important source for the creation of calligraphy, philosophy, literature, aesthetics and painting, contributing to the rich cultural achievements of great spiritual value.

1.3 Bearing the World in Mind, Contributing to Global Development

The natural resources and ecological environment in China are not only an important guarantee for sustainable development but also an important factor concerning global ecological safety, improvement and security. China has closely linked its development with that of the world and participates in international climate and environmental governance in a pragmatic way, upholding a global development philosophy of harmonious coexistence between humans and nature. The *United Nations Millennium Development Goals* (MDGs), formulated at the turn of this century, also embodied China's greatest efforts and contributions to the world, and the *2030 Agenda for Sustainable Development* (ASD), which was adopted at the UN Sustainable Development Summit 2015, clearly expressed the urgent need to "transform our world" and defined the development targets for "people, planet, prosperity, peace and partnership". In 2030 ASD, harmony with nature was an important theme, as it was emphasized several times, which coincided with the concepts of harmony between humans and nature and building a community with a shared future for mankind pursued by China. In China's 13th Five-Year Plan, China proposed the new development concepts of "innovation, harmonization, green, openness and sharing", which also received wide attention and appreciation from countries around the world. President Xi Jinping repeatedly stressed on different occasions that China would strive to build a modernization in which people and nature coexist in harmony. Therefore, focusing on the integrated governance of mountains, rivers, forests, fields, lakes, grasslands and deserts, governments at all levels in China and society as a whole have carried out a series of fundamental, groundbreaking and long-term projects. From national strategies, such as steadily promoting the construction of the Yangtze River Economic Belt, orderly advancing the protection of the Yellow River Basin and continuously fostering the high-quality development of 9 provinces along the Yellow River, to specific measures and policies, including carrying forwards environmental restoration projects such as returning marginal farmland into forests and grasslands, sparing no effort to carry out comprehensive control of desertification and soil erosion, and improving the public awareness of environmental protection, China assumes its due responsibility in various fields of propelling the theoretical explorations and practices of harmony between humans and nature and setting up examples and demonstrations for other countries around the world. After years of unremitting efforts, many cities in China have significantly improved air and water quality, as people can see blue skies and white clouds more often. In addition, forest coverage expands in China, the green economy develops faster, energy and resource consumption (per unit output) are decreasing, the number of days with heavy smog and haze is effectively reduced, the water quality of black and odorous water bodies is significantly improved, urban and rural environments are more liveable, and the construction of a beautiful China is taking solid steps. According to NASA satellite data, between 2000 and 2017, approximately a quarter of the new green area around world was in China. All these objective facts have proven China's practical efforts in promoting global green development, which was widely recognized and praised by the international community.

2 Explorations for Jointly Building a Shared Future for Humans and Nature

2.1 Joining Hands with Other Countries to Actively Combating Climate Change

Climate change has become one of the most significant challenges facing the world, posing a major threat to natural systems and the sustainable development of human society. Achieving the goal of building a community of harmony between humans and nature and addressing climate change requires the joint efforts of all countries and regions. Since the reform and opening-up, China has made remarkable achievements in economic development: between 2003 and 2013, the GDP increased nearly five times and per capita income grew nearly three times,[1] but the total energy consumption nearly quadrupled, and the total CO2 emissions nearly doubled.[2] In 2006, China topped Europe and the United States and became the world's top CO2 emitter. From then on, the share of China in Global CO2 emissions continued to climb year by year. The long-standing relatively extensive economic development model in China and its role as the "world factory" in the international division of labor have made it more urgent and harder for China to address climate change. To change the situation, the Chinese government adopted a series of policies and measures to effectively mitigate and adapt to climate change and made positive contributions to the world's efforts to combat climate change. Upholding the principles and spirit of the *United Nations Framework Convention on Climate Change* (UNFCCC), China has taken active measures to participate in and lead the global low-carbon transition. In June 2007, the Chinese government promulgated *China's National Climate Change Program* (CNCCP), outlining objectives, basic principles, key areas of actions, and policies and measures to address climate change and implement the UNFCCC. Since 2005, when China formulated the 11th Five-Year Plan, carbon emission intensity has been included as an important obligatory target in national economic and social development plans, ensuring a continuously reducing carbon emission intensity. Compared to 2005, the carbon emission intensity in 2018 decreased by 45.8%, meaning that China had realized the promised target of reducing carbon emission intensity by 40–45% by 2020 in comparison with that in 2005, three years ahead of schedule. China

[1] Calculated based on the *China Statistical Yearbook* over the years.

[2] Based on statistics from the *BP Statistical Review of World Energy* over the years.

basically reversed the momentum of the rapid growth of greenhouse gas emissions and generally improved the situation of greenhouse gas emissions.

In addition, China is actively taking part in global climate governance. In 2017, President Xi Jinping emphasized in his speech delivered at the opening ceremony of the World Economic Forum Annual Meeting that the *Paris Agreement* was in line with the general direction of global development and that countries should uphold the hard-won result and should not give up lightly. The speech made historical contributions to the conclusion of the *Paris Agreement* and greatly encouraged the world to jointly address climate change. In 2020, President Xi Jinping emphasized in his speech at the Leaders' Side Event on Safeguarding the Planet of the G20 Riyadh Summit that "the Earth is our shared home. We may follow the vision of a community with a shared future for mankind and pitch in to tackle climate and other environmental challenges and protect the blue planet". In 2022, President Xi Jinping replied to a letter from British primary school students on the issue of climate change, saying that the Earth is a big family, and the human race belongs to one community and that as climate change poses a common challenge to all humanity, mankind should cooperate to tackle the issue. In September 2020, President Xi Jinping delivered an important speech at the General Debate of the 75th Session of the United Nations General Assembly, in which he put forwards the vision that China "aims to have carbon dioxide emissions peak before 2030 and achieve carbon neutrality before 2060", demonstrating the determination of China to give a lead in promoting the construction of a global climate governance system for the common interests of China and the world and the common welfare of all mankind. As China set carbon neutrality before 2060 as a clear development goal, the other countries were encouraged. The EU raised its climate target of the reduction in carbon emissions again, from 55 to 60%, compared to 1990. Japan, South Korea, the United Kingdom and other countries have also put forwards their own goals of achieving carbon neutrality by 2050. The example of China received a positive response globally, having a due demonstration and leading effect. As UN Secretary-General Guterres said, when he offered congratulations on the Communist Party of China's centennial birthday to come and the 50th anniversary of the restoration of the PRC's lawful seat in the United Nations, the United Nations thanked China for "upholding multilateralism, and supporting the UN's work, and highly appreciate China's declared targets of intended nationally determined contributions and major initiatives to combat global climate change".

2.2 Strengthening Regional Cooperation, Constructing a Green Belt and Road Through Consultation and Joint Contribution

The gap between the rich and the poor has widened due to the overlapping of various factors, including the epidemic, slowing economic growth, increasingly complex geopolitics, and the increasingly prominent contradiction between economic and social development and eco-environmental constraints, since balancing the relations between human society and natural ecology is still a global challenge in the new century. As comprehensive national power grows and cooperation with the rest of the world intensifies, China is playing an increasingly important role in global governance. The Belt and Road Initiative provides a new option to solve the contradiction between development and the environment, reflecting China's taking on its responsibilities and contributing to human social development and global governance as a responsible power.

As of August 2018, China had signed 118 Belt and Road cooperation agreements with 103 countries and international organizations, and the joint building of the Belt and Road and the principles of consultation, joint contribution and sharing had been written into the outcome documents of important international institutions, including the UN.[3] China has strengthened trade and investment in countries along the Belt and Road, boosting local employment and economic development. According to a spokesman for the Ministry of Commerce of China, since China launched the construction of the Belt and Road, Chinese enterprises have built 75 overseas economic and trade cooperation zones in countries along the Belt and Road, with a cumulative investment of 25.5 billion U.S. dollars, paying nearly 1.7 billion U.S. dollars in taxes and fees to the host countries and creating nearly 220,000 jobs.[4]

In addition, China is committed to green development. In 2017 and 2019, General Secretary Xi Jinping emphasized in his speeches at the First and Second Belt and Road Forum for International Cooperation that "we should pursue the new vision of green development and a way of life and work that is green, low-carbon, circular and sustainable" and that China "aims to promote green development, and may launch green infrastructure projects, make green investment, and provide green financing". In promoting the construction of the Belt and Road, especially in developing countries, China focuses on the implementation of infrastructure projects for public welfare, people's livelihood and energy and makes full use of China's concepts, experience, technology and equipment of green and low-carbon development to develop energy-saving and environmentally friendly solutions for local communities and maximize the application of smart, green and low-carbon technologies in the projects. When constructing large aid projects such as the opera house in Algeria and the wrestling arena in Senegal, Chinese contractors took the initiative to introduce green building and clean energy technologies to effectively improve energy consumption and reduce environmental pollution due to construction. When carrying out key municipal projects such as the water supply and sewage treatment project in Kurunegala, Sri Lanka, and the road renovation and community drainage project in Sao Tome and Principe, Chinese contractors solved the problems of waterlogging and

---

[3] Lu Ya'nan, "Belt and Road: More Friends, More Opportunities" (at the press conference of the State Council Information Office), http://politics.people.com.cn/n1/2018/0828/c1001-30254367.html.

[4] Global Times (Finance),"The Belt and Road Initiative May Boost Global Trade to Grow by 12%: Cutting Trade Costs Between Countries", http://finance.huanqiu.com/chanjing/2018-06/12202090.html?agt=62.

rain and sewage diversion quickly and efficiently, effectively improving the quality of life of local residents. Through the Belt and Road Initiative, China has promoted the development of the local clean energy industry with infrastructure projects, including hydropower, solar, wind, nuclear and geothermal power. For example, the installed capacity of the solar power station aid project in Cuba has reached nine megawatts, promoting energy independence, energy security and the industrialization of related industries in Cuba, as well as local economic development. Relying on the Belt and Road Initiative, China carries out green investment and green construction projects in countries along the Belt and Road, simultaneously improving the local natural and human environment and providing local residents with a more liveable environment.

2.3 Advocating a Community with a Shared Future for Mankind and Jointly Charting the Course for Global Ecological Civilization

The ecological civilization proposed by China skips the development stage of industrial civilizations and goes beyond the traditional scope of energy saving, emission reduction, resource conservation and environmental protection. It advocates a fundamental transformation in the social development model and, from the top, redefines and regulates the relationship between humans and nature, between individuals and society, and among individuals in terms of social value orientation and human civilization progress. At present, China is vigorously advancing ecological progress, which not only aims at the phased goals of promoting economic growth and alleviating pressure on resources and the environment but also targets the grand vision of guiding social development with advanced concepts, ideas and culture. In the progress of governance and social development, China attaches the same importance to ecological progress and other national strategies and national sustainable development. Practicing ecological civilization, insisting on green development and building a beautiful China are all important elements in realizing the Chinese dream of the grand rejuvenation of the Chinese nation, as well as the efforts and commitments China makes to global development. China is building an ecological civilization that upholds the concept of "a community with a shared future for mankind". As the concept indicates, in the process of globalization, human beings belong to a community that shares both prosperity and loss. In February 2013, China's ecological civilization vision was formally included in the decision documents of the 27th Governing Council of the United Nations Environment Programme. The innovative concepts and practical experience of China in ecological civilization construction have been widely recognized by the international community and have attracted increasing attention around the world. From the perspective of global development, by implementing the national strategy of ecological civilization construction, China has achieved the comprehensive integration of actual economic and social development with the concept of sustainable global development, stimulating theoretical explorations and concrete practices in relevant fields and positively promoting the carrying-on, innovation and development of the concepts of sustainable global development.

3 Chinese Experience in Jointly Building a Shared Future for Humans and Nature

3.1 Adhering to the Harmony Between Humans and Nature and Building an Ecologically Livable World

Since the 18th CPC National Congress, adhering to the concept of harmony between humans and nature, China has comprehensively improved the quality of the ecological environment and made great historical achievements in the green and low-carbon transformation and development of the whole society and various industries. As the Third National Land Survey indicated, between 2009 and 2019, the area of woodlands, grasslands, wetlands, rivers, lakes and other types of land with strong ecological functions in China increased by 260 million *mu* (42.83 million acres). The area of desertified and sanded land continued to shrink. In 2020, the national forest coverage rate rose to 23.04%, and the forest stock volume increased to 17.56 billion cubic meters. The populations of some endangered species remained stable or grew slightly. The proportion of surface water bodies nationwide rated good water quality increased from 61.7% in 2012 to 84.9% in 2021, the proportion of days with good air quality in cities at the prefecture level or above rose by 6.3 percentage points in 2021 compared to 2015, and carbon emissions per unit of GDP decreased by 48.4% in 2020 compared to 2005. These figures directly demonstrate the determination of the Chinese government and society as a whole to protect the environment and transform to green development and clearly indicate the achievement and practical effects of China's efforts to build a new ecologically liveable world. There are many examples in China showing that both ecological livability and development are achievable at the same time.

Guangyuan City in Sichuan Province is located in the Qinling-Daba Mountainous Region, and there used to be a weak industrial base and difficult living conditions, with per capita GDP only approximately 30% of the national average, and three national poverty-stricken counties, Cangxi County, Wangcang County and Chaotian District. In the face of such a huge pressure on developing the economy and improving people's lives, Guangyuan municipal government chose another path of development bravely: it insisted on building an ecological city and developing an ecological economy, discarding the old model of expanding traditional industries like other cities did in the early stages of industrialization and urbanization, and set up a model for ecological civilization construction.

Guangyuan strictly adheres to the ecological red lines, gives priority to the conservation of "clear waters and lush mountains" there and has taken active protective measures. For example, Guangyuan spent more than USD 110.72 million on the comprehensive treatment of 19 urban black and odorous water bodies and constructed sewage treatment plants in all counties and towns along the rivers, realizing effective treatment of domestic sewage in 65% of administrative villages in Guangyuan and turning the Nanhe River and other urban black and odorous water bodies into lucid waters. In addition, Guangyuan invested more than USD 13.84 million to sort out 14,700 fish cages in the Bailong River, an important tributary of the Jialing River, to improve the water quality of the Bailong River, raising the water quality rating of the Bailong River to the Grade I standard, and the water quality of the Guangyuan section of the Jialing River to Grade II. "A river of clear water flows out of Guangyuan" has become a common ecological saying in Guangyuan. With comprehensive measures to improve water, air and soil quality, the number of days with good air quality in the main urban areas in Guangyuan has topped 350, 95.24% of arable land has received national pollution-free certification, and the forest coverage rate has risen to 56.18%. Guangyuan has built a solid ecological barrier in the upper reaches of the Jialing River.

Bailong River in Guangyuan

Source Bailong Lake Scenic Park Administrative Bureau of Guangyuan City, Sichuan Province, "Three Provinces Will Establish a Coordinated Prevention and Control System for the Protection and Governance of Bailong Lake", http://blhglj.cngy.gov.cn/bailonghu/article.html?id=1310

To coordinate industrial development with the environment, Guangyuan City gives full play to its characteristic advantages in location and resources and develops a modern ecological industrial system by insisting that industrial development should accommodate, take base on and improve the ecology. Starting from the investment attraction, Guangyuan City rejected high pollution and high energy consumption enterprises and carried out strict environmental assessments for postdisaster reconstruction projects in Wenchuan, issuing no approval or construction licence to the project or enterprise failing the environmental assessment. In addition, Guangyuan determinatively and comprehensively eliminated the backwards production capacity in cement, ferroalloy, coke, coal mining and other high-polluting industries and resolutely inhibited the development of graphene, coal mining and other mining industries while strengthening support for new materials, new energy, new pharmacy, agricultural and sideline product processing and other emerging industries.

Relying on the ecological background and green endowment, focusing on modern agriculture and targeting increasing farmers' income, Guangyuan City made special regional development plans for local resources and developed into an important production base for Chinese herbal medicine, tea, specialty fruits, oil olive and high-quality edible mushrooms in China. Specifically, Jian'ge County in Guangyuan City is now an important oil plant planting county, Cangxi County is a national production base for snow pears and kiwis, and Lizhou District is a national anti-seasonal vegetable planting base, with the production of chuanminshenviolaceum, radix bupleuri and rhizomagastrodiae (traditional Chinese herbal medicinal materials), ranking first in Sichuan Province and among the top herbs in China. Thanks to the rapidly advancing agricultural scale-production, standardization and industrialization, Guangyuan has become well-known in China as a planting base of snow pear and specialty chuanminshen violaceum, a construction demonstration base of walnut, a production base of tu-chung, and the only and the largest green aquatic product demonstration zone of green product demonstration zones in western China featuring salangid and longsnout catfish, initially forming an industrial structure enabling high-quality regional development based on fruits, tea, vegetables, herbs, grain and oil, tobacco and other leading agricultural industries.

Supported by its advantageous resources, Guangyuan City is now making greater efforts in the ecological industry to constantly develop new and strengthen existing industrial advantages, boost green, low-carbon and circular growth, foster its own low-carbon characteristic industries, create brands of Guangyuan City that integrate medical and health care and tourism, and establish a new image and development platform for the city. Therefore, Guangyuan City allows full play of the wisdom of the government and the initiative of the people to achieve the optimal allocation of ecological assets, provide ecological products with "excellent quality and reasonable prices", and diversify ecological products and industries in different counties and townships, and even in villages. Meanwhile, with subsidies from "Beautiful Sichuan and Livable Countryside" construction funds, people-benefit green development campaigns, integrated ecological product commercialization and rural revitalization and other initiatives, Guangyuan City is improving both the local eco-environment and living conditions.

The experience of Guangyuan City proves that the harmony between humans and nature is the basis of long-term development, that social development and urban construction must follow the laws of nature and that the only way to truly become prosperous and sustainable is to protect and reasonably use natural resources based on local conditions and establish a green, low-carbon and sustainable mode of production and consumption.

3.2 Upholding Green and Low-Carbon Development, Building a Clean and Beautiful World

The rapid development of the global economy relies heavily on and benefits from the exploitation of fossil energy on a large scale, which not only brings climate and environmental problems but also poses certain threats to biosecurity, ecological balance and human survival while promoting economic development. In recent years, China has focused on new energy sources to advance the joint construction of a clean and beautiful world. Since 2012, China has made outstanding achievements in green and low-carbon energy transition and energy structure optimization. According to the statistics of the National Energy Administration, China's renewable energy has achieved leapfrog development, with an installed capacity exceeding 1.063 billion kilowatts, accounting for 44% of the total installed power generation capacity in China. Specifically, the installed capacity of hydroelectric power has been holding first place around world for seventeen consecutive years and that of wind, photovoltaic and biomass power for twelve, seven and four consecutive years, respectively, raising the proportion of clean energy in energy consumption to 25.5%, from 14.5%. Meanwhile, wind power, photovoltaics, energy storage, nuclear power, extrahigh voltage and other sectors are advancing rapidly in China, laying a solid technical and industrial foundation for the greening and upgrading of energy.

Qinghai Golmud Solar Park is the world's largest solar park and a typical example of green energy development in China. In Golmud, there used to be "only yellow sand but no green trees", but through the construction of photovoltaic parks and the promotion of afforestation, this region has planted forests of more than 9,700 *mu* (1,598 acres) in total. At the same time, Golmud applies high standards to construct ecological and green photovoltaic parks to transform environmental advantages into economic benefits, achieving balances among eco-environmental protection, economic development and improvement of life quality. The development of the new energy industry in Golmud is just a microcosm of Qinghai Province, reflecting the booming clean energy industry there from one aspect. Qinghai set a new world record for a full clean energy power supply in 2017 and 2019 with the "Seven Days of Green Power" and "Nine Days of Green Power" campaigns, respectively. From June 9th to 23rd, 2019, all electricity consumption in Qinghai Province came from water, solar, wind and other clean energy sources, making a record of full clean energy power supply for 360 consecutive hours for a province and zero emission for 15 days. In accordance with the target orientation of carbon peaking and carbon neutrality, the western regions of China will fully implement strategic plans for new energy by planning and building large-scale wind and photovoltaic power bases of 450 million kilowatts in areas with poor ecological resources. The projects under construction have an installed capacity of 85 million kilowatts. The energy structure, ecological environment and basic economic conditions in western China will be further improved and optimized.

Energy storage power station in Golmud, Qinghai Province

Source Xinhua News Agency, "Golmud in Qinghai: Building Momentums on Photovoltaic Panels", https://www.sohu.com/a/342239988_267106?g=0&spm=smpc.content.share.1.1591096959958zvyroeu

In the future, China will accelerate the development of green and clean energy and renewable energy, promote the sustainable use of natural resources, and advance the planning and construction of large-scale wind power and photovoltaic projects in deserts, Gobi and desert areas based on local ecological conditions and priority development strategies to actively address climate change and comprehensively carry forwards the cause of ecological conservation.

3.3 Building an Open and Inclusive World Through Exchanges and Mutual Understanding

President Xi Jinping repeatedly stressed on different occasions that the reform of the global governance system must be led by suitable concepts, and the innovative development of global governance concepts should fully explore the positive meanings and essence of the philosophy of life and governance of traditional Chinese culture. On

November 30th, 2015, Chinese President Xi Jinping delivered an important speech entitled *Work Together to Build a Win–Win, Equitable and Balanced Governance Mechanism on Climate Change* at the opening ceremony of the Paris Conference on Climate Change, in which he specified the global governance concepts of "win–win cooperation, with each country making contribution to the best of its ability", "the rule of law, fairness and justice" and "inclusiveness, mutual learning and common development", while he advocated the synergistic approach to development of "harmony without uniformity" to allow countries to seek their own solutions that best suit their respective national conditions. Based on Chinese culture and wisdom, these concepts are in keeping with the aspirations of developing countries and coincide with the development trend of global integration.

In 2015, during a visit to South Africa for the Johannesburg Summit of the Forum on China-Africa Cooperation (FOCAC), Chinese President Xi Jinping suggested that China-Africa cooperation should strive to accelerate the industrialization and modernization of agriculture in Africa and that China would help Africa start the industrialization plan quickly. Located in Dukem City, with a planned area of 5 square kilometers, the Ethiopia Eastern Industry Zone is a model for China-Africa international cooperation, south–south cooperation, the Belt and Road Initiative and overseas economic and trade cooperation zones. During construction, China has always upheld the principles of equality, friendship, consultation and win–win cooperation to promote the sustainable construction and development of the industrial zone based on actual conditions such as rich human resources and resource endowment in Ethiopia. To date, the industrial zone has attracted more than 100 enterprises and created more than 20,000 job opportunities for local employees. As a priority project of the industrial development plan, the Ethiopia Eastern Industry Zone has been listed as a partner of the "Sustainable Development and Poverty Reduction Program" in Ethiopia. In recent years, China introduced and accumulated rich green and low-carbon development experiences in the construction, operation and management of the industrial zone, and the results have effectively helped improve the ecological, economic and social environment in Ethiopia, providing a reference for accelerating the international implementation of ecological civilization theories. Taking the Eastern Industry Park in Ethiopia as a pilot and model, China is continuously perfecting systematic support such as technical exchange, legal consultation, medical security and education and enhancing comprehensive guarantee and soft support capabilities in fields such as law, technology, talents, economy and the people's livelihood, to develop new models, accumulate new experience and set up new examples for China-Africa cooperation and international cooperation. By doing so, China is effectively opening up new channels for China-Africa exchanges and forming new stimuli for multilateral and diversified cooperation on green development.

Ethiopia Eastern Industry Zone

Source Eastern Industry Zone, "Images of the Zone", http://www.e-eiz.com/img1.php

For decades, the Chinese government and enterprises have adhered to the principles of mutual trust and assistance, equality and consultation, and win–win cooperation and have formed a solid community with a shared future with African countries through close cooperation in a complex global environment, facing difficulties together, progressing hand in hand and sharing achievements. In 2018, the Beijing Summit of the Forum on China-Africa Cooperation was held, and China and Africa unanimously decided to explore and advance the building of a closer China-Africa community with a shared future to assume joint responsibility, pursue win–win cooperation, deliver happiness for all, enjoy cultural prosperity, ensure common security and promote harmony between humans and nature, setting a model for furthering the building of a community with a shared future for mankind.

Under the new international situation of increasingly complex geopolitics and the persistent epidemic of COVID-19, China will strengthen its pragmatic cooperation with all countries and regions in the world, especially developing countries, to jointly and actively solve environmental and development problems with all parties, improve mutual learning between civilizations, and deepen partnerships, striving to build a higher quality, more efficient, more equitable and more sustainable development mechanism. China will play the role of a major power, forge global consensus, and work with all countries and parties in the world to shoulder the historical mission of addressing global development challenges, cultivating new momentum for global development, and building an inclusive global development governance structure.

GPSR Compliance

The European Union's (EU) General Product Safety Regulation (GPSR) is a set of rules that requires consumer products to be safe and our obligations to ensure this.

If you have any concerns about our products, you can contact us on ProductSafety@springernature.com

In case Publisher is established outside the EU, the EU authorized representative is:

Springer Nature Customer Service Center GmbH
Europaplatz 3
69115 Heidelberg, Germany

Batch number: 10756378

Printed by Printforce, the Netherlands